[개정판]

미용인을 위한 실무영어

Business English for Beauticians

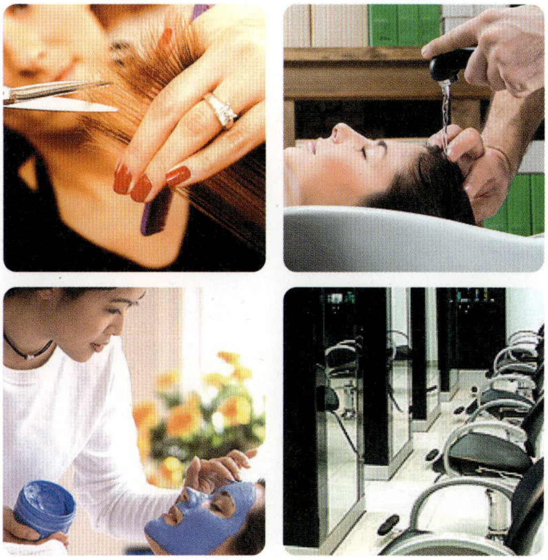

김기영 · 신규옥 · 안경민 · 김수미 · 송다해 · 최은미

메디시인

머리말

21세기는 글로벌화가 되면서 전 세계는 영어라는 공용어의 사용과 필요성이 매우 높아지고 있는 실정이다. 이러한 추세는 미용 분야에서도 예외일 수는 없는 상황에 있고, 아름다움의 추구라는 하나의 패러다임 안에서 미용학은 학문으로서 자리매김을 확실하게 하고 있다. 글로벌화가 되면서 국가 간의 교류가 많아지고 우리나라의 미용은 한류와 함께 k-Beauty는 전 세계적으로 그 영역과 활동 범위가 확대되면서 미용인들의 실용 영어의 필요성이 그 어느 때보다도 높아지고 있다.

따라서 본서는 미용에 입문하려는 미용인들에게 있어서 필수적으로 알아두어야 할 분야인 헤어, 피부 관리, 메이크업, 네일 등과 관련된 미용 용어와 기초 영어 회화에 대한 핵심적인 내용을 간략하게 전달하고자 집필하였다.

교재의 전체적인 구성은 Part Ⅰ, Ⅱ, Ⅲ로 나누었으며 Part Ⅰ에서는 미용 기초 관련 용어, Part Ⅱ에서는 고객 전화 상담과 방문에 대한 내용을 Hair, Esthetics, Makeup, Nail 분야, Part Ⅲ에서는 Hair, Esthetics, Makeup, Nail 분야의 관리와 조언을 구분하여 실제적으로 적용할 수 있도록 구성하였으며, 부록에서는 미용에 필요한 색표와 미용과 해부학 용어를 수록하였다. 본 서의 자료는 Milady's Cosmetology, 비달사순, 토니 앤 가이, 일부 산업현장에서 활용되고 있는 영어표현을 위주로 구성하여 미용인이라면 누구나 반드시 갖추어야 하는 미용 영어들을 체계적으로 정리하였으며, 이번의 출간을 계기로 미용 영어에 대한 표준의 교육지침이 마련될 수 있는 기틀이 되기 바란다. 미용 관련 물품, 도구와 기기는 인터넷 관련 사이트의 자료를 활용하였고, 미용 관련 서적에서 많이 사용되는 용어를 위주로 하였다.

끝으로 이 책의 출간을 위해 도움을 주신 임 연옥박사님, 그리고 지구문화사 편집부원들의 수고에 깊은 감사를 드립니다.

2016년 9월

저자 일동

차례

PART I Basic English in Beauty

PART II Telephone Reservation & Business Hours Inquiry

Chapter 01 Hair & Scalp Care

Chapter 02 Skin care

Chapter 03 Makeup

PART Ⅲ Care & Consultation

Chapter 01 Hair

Chapter 02 Skin Care

Chapter 03 Body Care

Chapter 04 Makeup Practice

Chapter 05 Nail Consultation & Care

Appendix 부록

Part I

Basic English in Beauty

Business English for Beauticians

1. Human Body

인체의 명칭

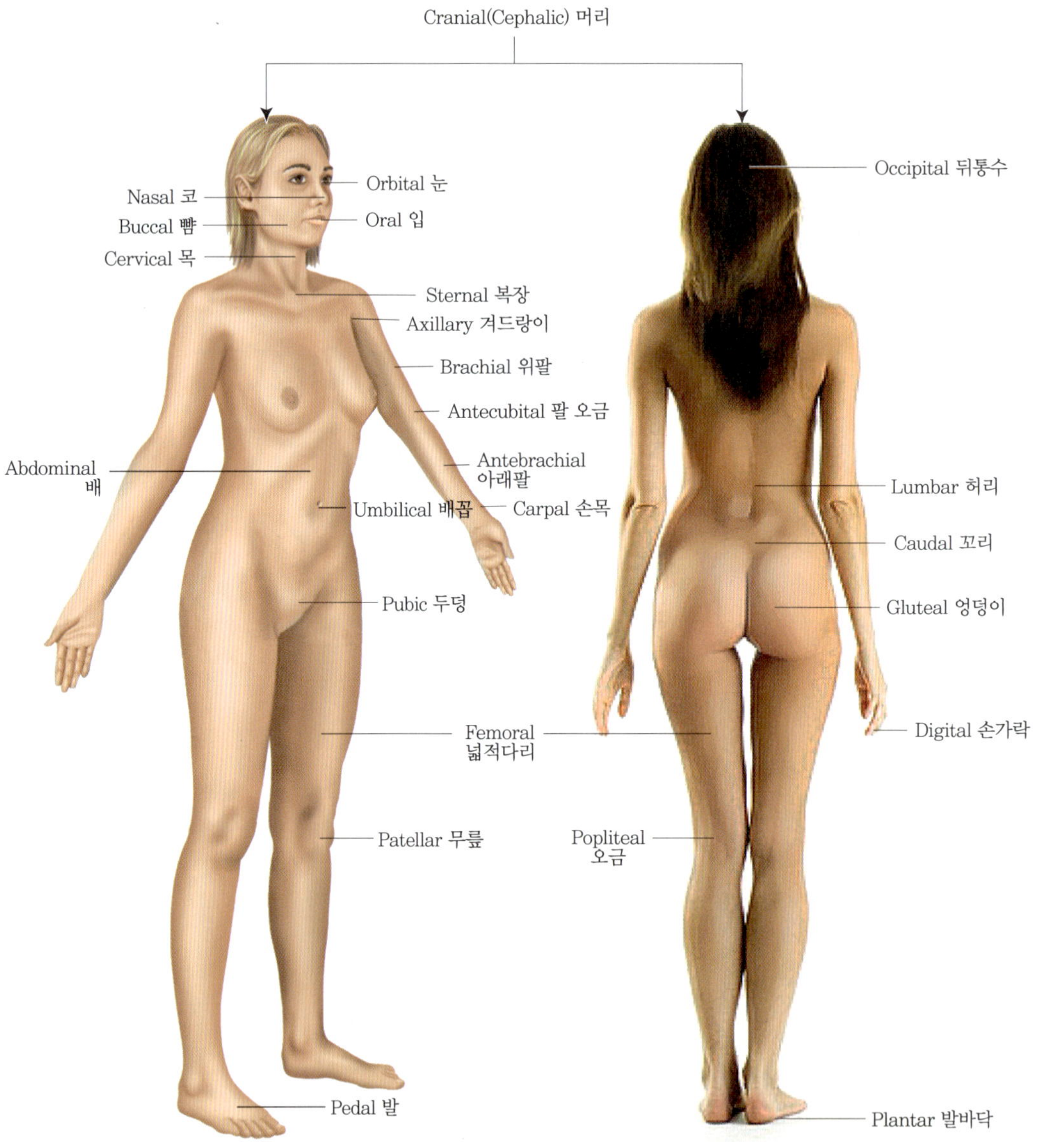

얼굴의 명칭

2. Structure of Hair, Skin and Nail

Hair Sturcture (헤어 구조)

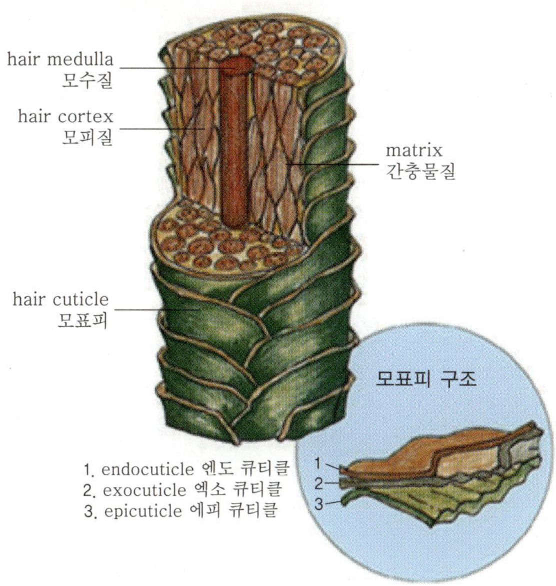

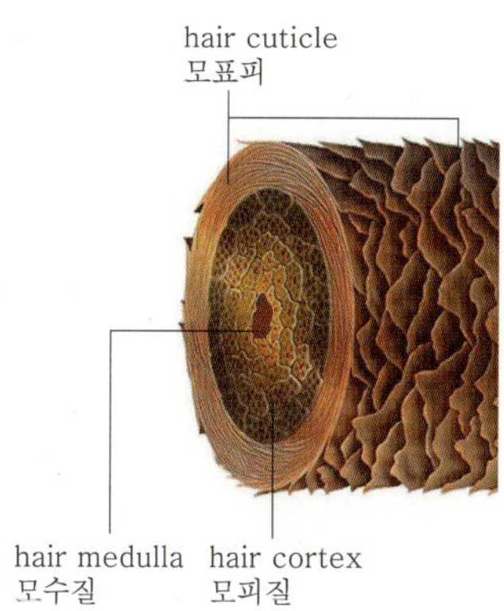

Skin Sturcture (피부 구조)

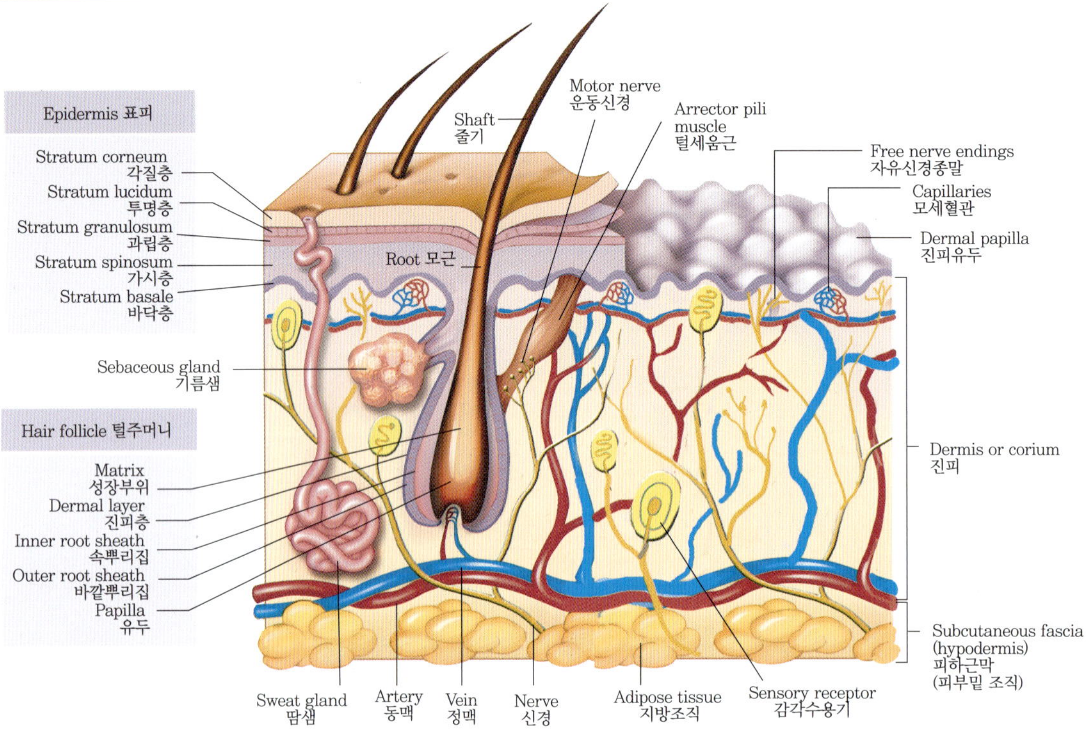

Nail Sturcture (조갑구조)

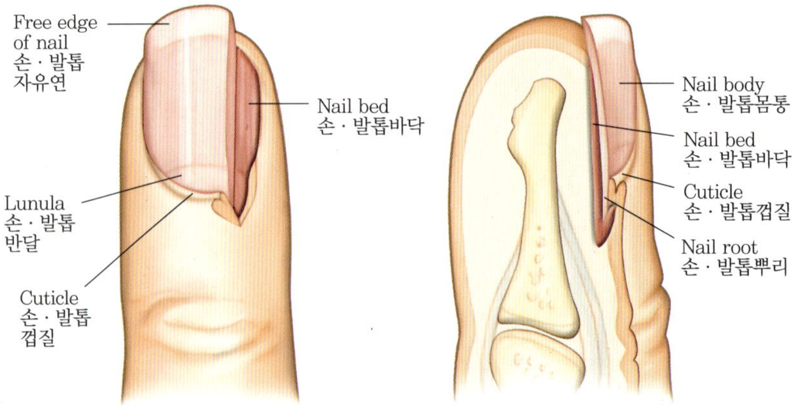

3. Materials & Tools for Hair

Rod	Comb	Brush	Tail Comb
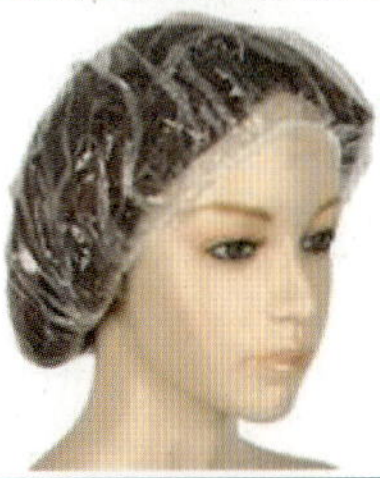		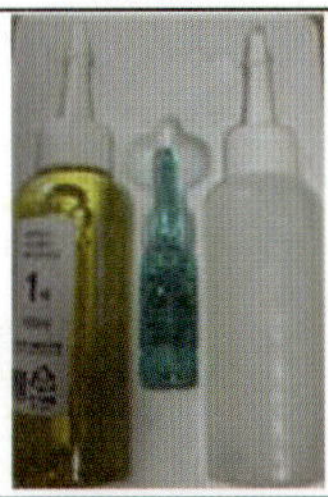	
Disposable Hair Cap	Pin Set	Oxidizer & Deoxidizer	Hair spray
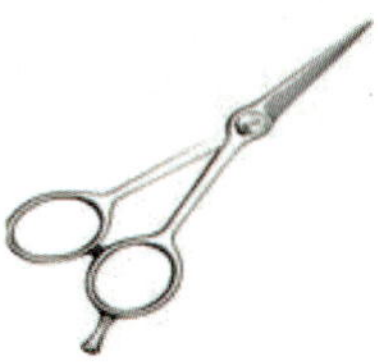		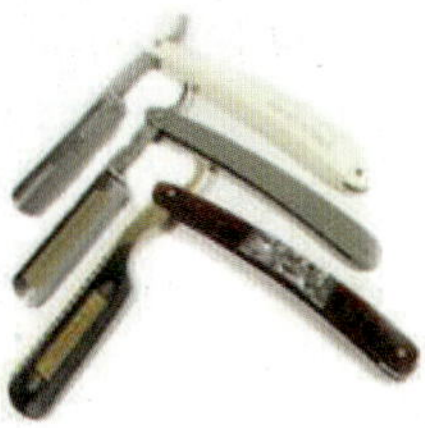	
Mini Scissors	Cutting & Thinning Scissors	Razor	Iron
		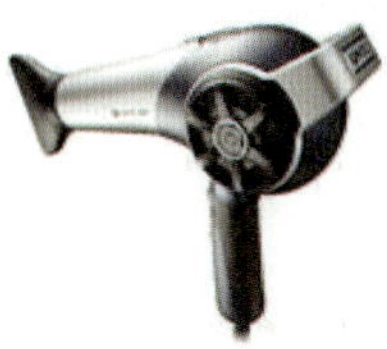	
Hair steamer	Clipper	Dryer	Hair Roll

4. Hair Accessories

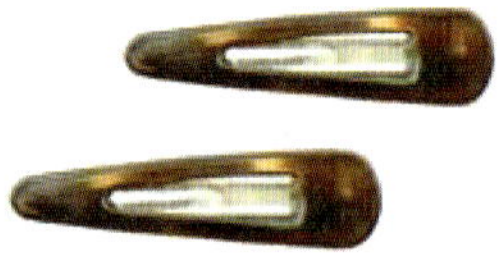 Clip	 Barrette	 Scrunchie	 Zigzag Headband
 Tiara	 Brooch Pin	 Elastic Band	 Cameo Ponytail Holder
 Snood	 Bobby pin	 Comb	 Claw Cip
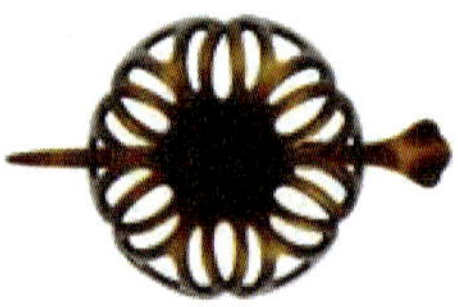 Bun Cage	 Bun Slide	 Bracelet Cuff	 Bracelet
 Hair Stick		 Chignon Pin	

5. Structure of scissor

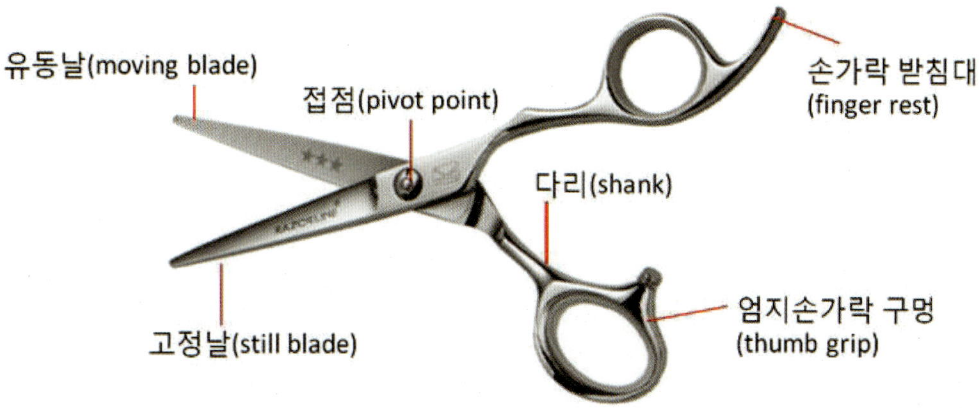

6. Shape of Face

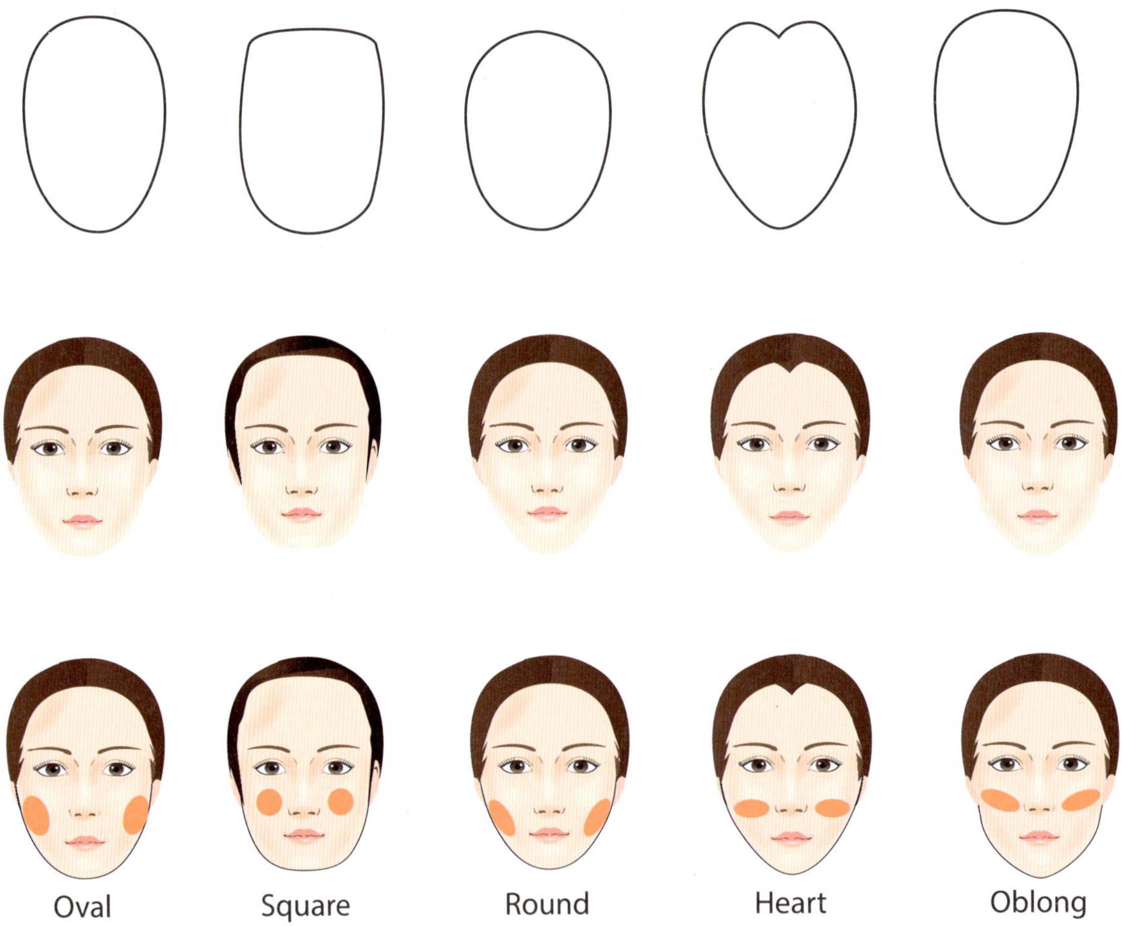

Hourglass

Inverted Triangle

Rectangle

Triangle

Diamond

Oval

7. Name of Hairstyles

Name	Hair Style	Name	Hair Style
Afro		Asymmetric cut	
Beehive		Blunt cut	
Bob cut		Blunt cut	

Name	Hair Style	Name	Hair Style
Bowl cut	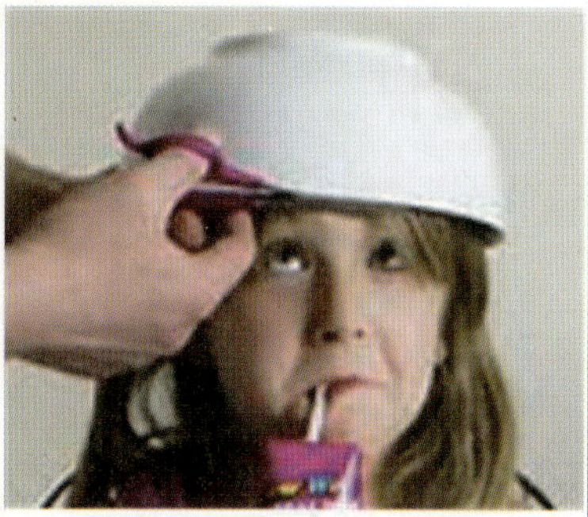	Braid	
Bun		Butch cut	
Buzz cut	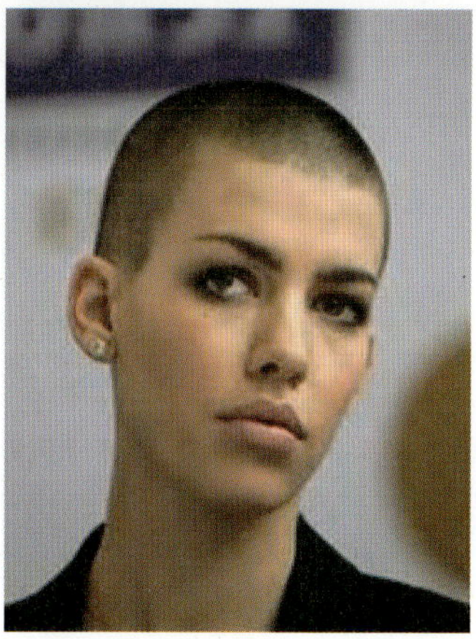	Caesar cut	

Name	Hair Style	Name	Hair Style
Chignon		Chonmage	
Comb over		Conk	
Cornrows		Cropped hair	
Crown braid		Curtained hair	

Name	Hair Style	Name	Hair Style
Devilock		Dreadlocks	
Fallera hairdo		Feathered hair	
Finger wave		Fishtail hair	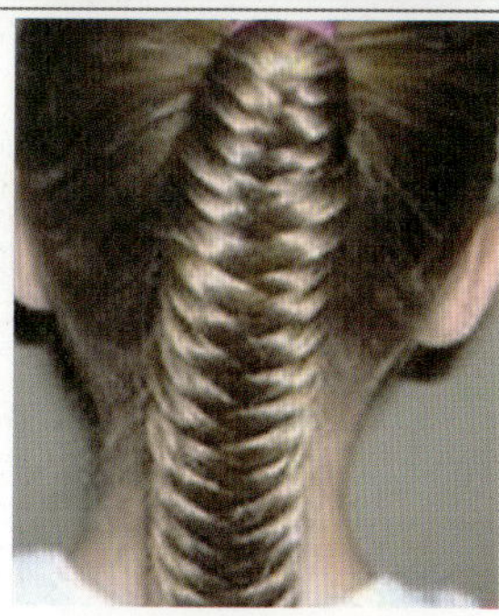
Fontange		French braid	

Name	Hair Style	Name	Hair Style
French twist	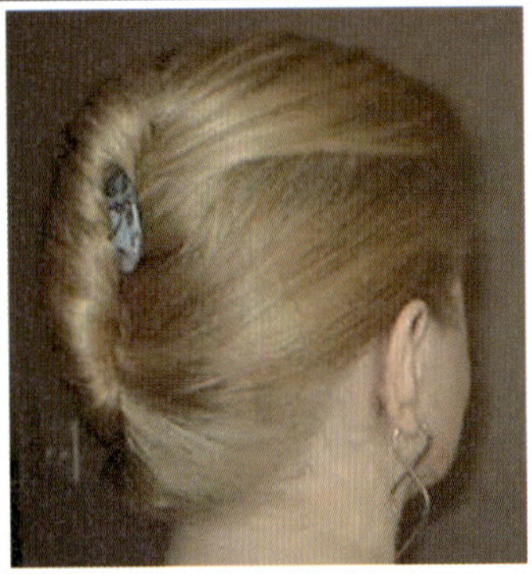	Fringe(bangs)	
Hime cut		Hi-top fade	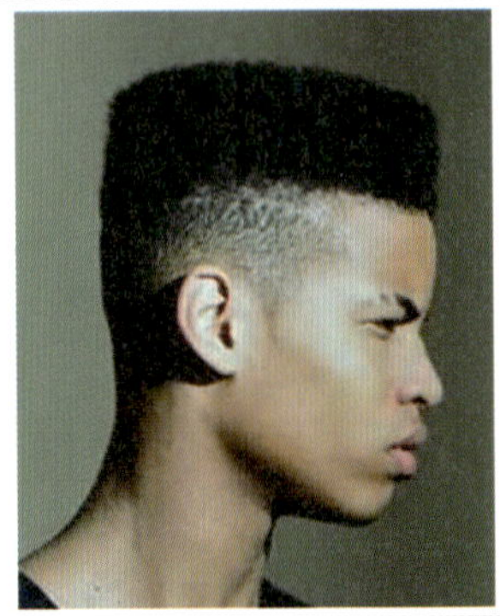
Jewfro		Liberty spikes	
Marcelling		Mohawk	

Name	Hair Style	Name	Hair Style
Mop top		Mullet	
Odango		Pageboy	
Payot		Pigtails	
Pixie cut		Pompadour	

Name	Hair Style	Name	Hair Style
Ponytail		Psychobilly Wedge	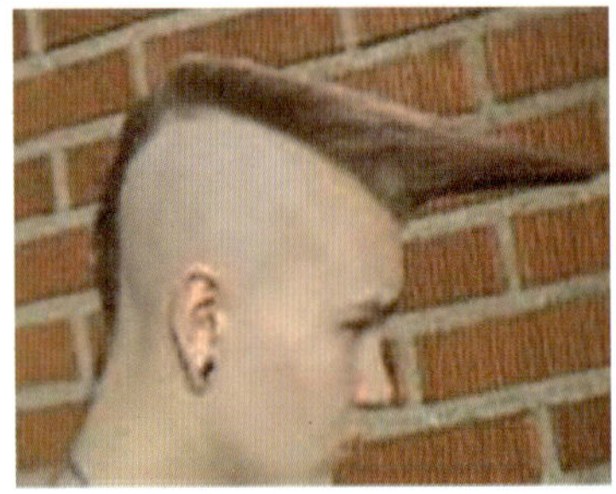
Rattail	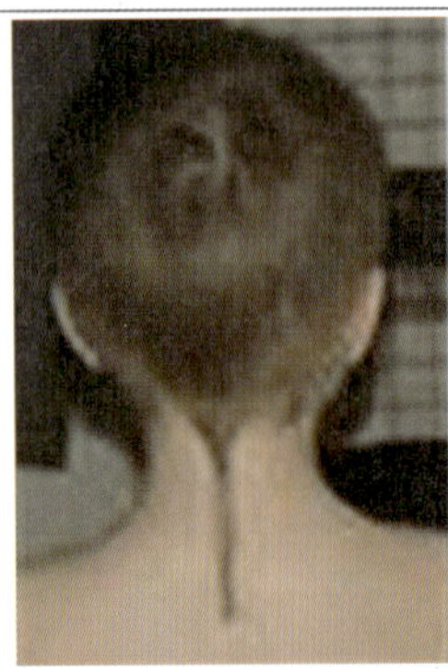	Ringlets	
Shag		Spiky hair	
Tonsure	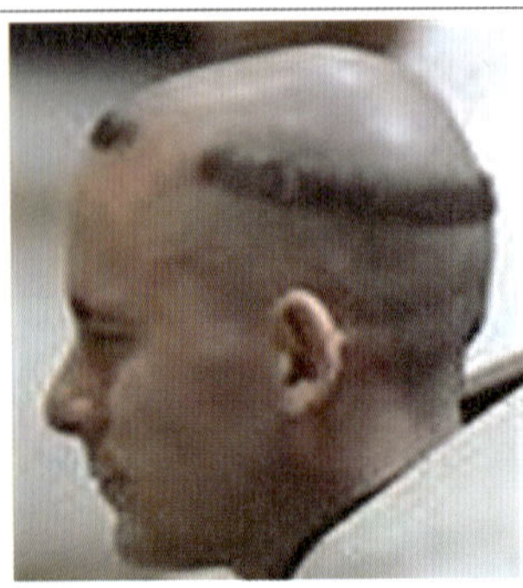	Updo	

8. Hair Section Point

위치	약자	명 칭
1	E.P	이어 포인트(Ear Point)
2	C.P	센타 포인트(Center Point)
3	T.P	탑 포인트(Top Point)
4	G.P	골든 포인트(Golden Point)
5	B.P	백 포인트(Back Point)
6	N.P	네이프 포인트(Nape Point)
7	F.S.P	프론트 사이드 포인트(Front Side Point)(left/right)
8	S.P	사이드 포인트(Side Point)(left/right)
9	S.C.P	사이드 코너 포인트(Side Corner Point)(left/right)
10	E.B.P	이어 백 포인트(Ear Back Point)(left/right)
11	N.S.P	네이프 사이드 포인트(Nape Side Point)
12	C.T.M.P	센타 톱 미디움 포인트(Center Top Medium Point)
13	T.G.M.P	톱 골덴 미디움 포인트(Top Golden Medium Point)
14	G.B.M.P	골덴 백 미디움 포인트(Golden Back Medium Point)
15	B.N.M.P	백 네이프 미디움 포인트(Back Nape Medium Point)

9. Materials & Tools for Skin Care

	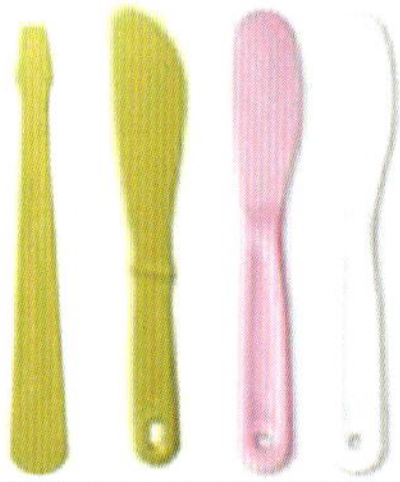	
Sponge	Spatula	Pack Bowl
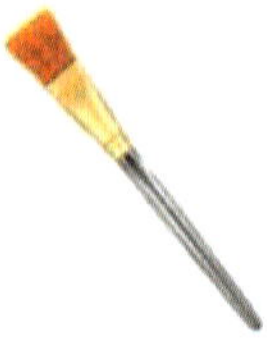	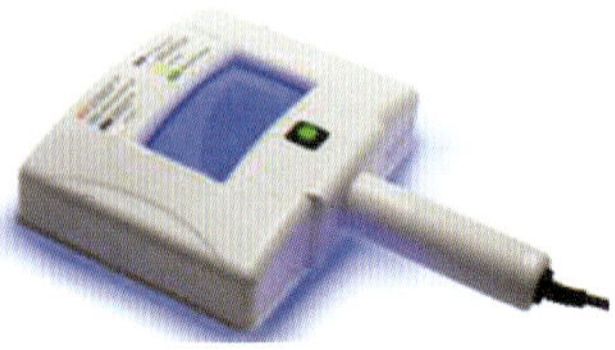	
Brush	Wood's Lamp	UV Sterilizer
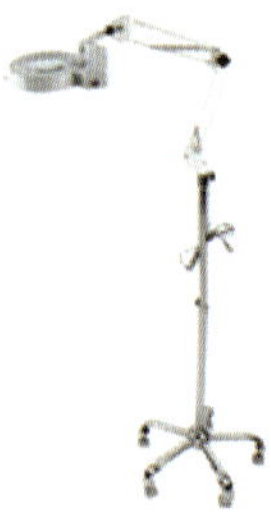		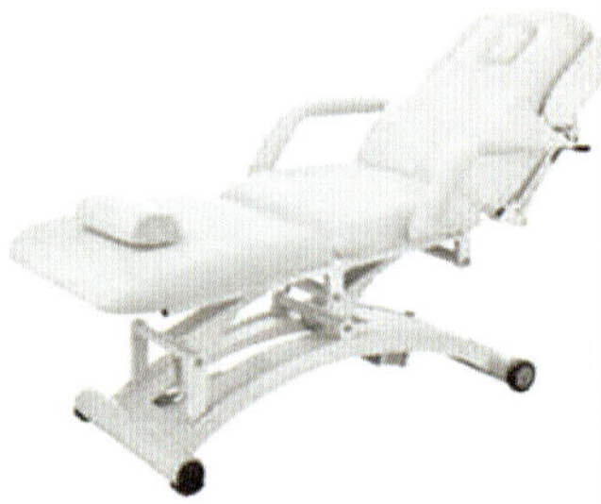
Magnifying Glass	Heating Cabinet	Automatic Bed
	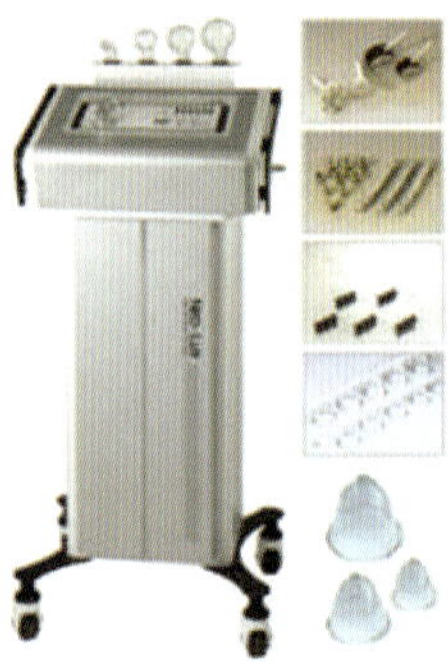	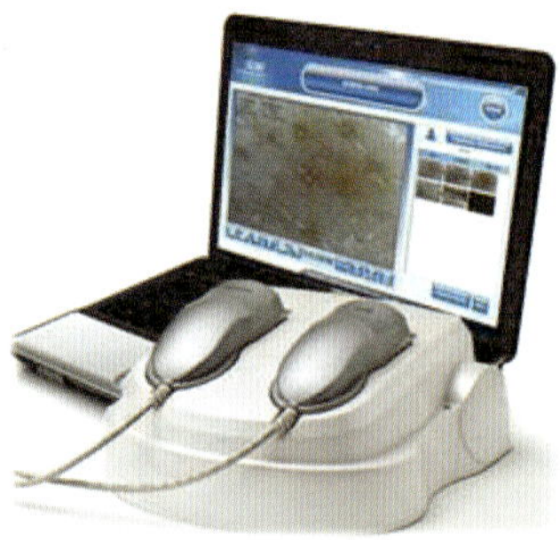
Steamer	Vacuum Suction	Skin Analyzer

10. 12 Meridians

번호	경맥 한글 이름	약 자	영 어
1	수태음폐경	LU	Lung Meridian
2	수양명대장경	LI	Large Intestine Meridian
3	족양명위경	ST	Stomach Meridian
4	족태음비경	SP	Spleen Meridian
5	수소음심경	HT	Heart Meridian
6	수태양소장경	SI	Small Intestine Meridian
7	족태양방광경	BL	Bladder Meridian
8	족소음신경	KI	Kidney Meridian
9	수궐음심포경	PC	Pericardium Meridian
10	수소양삼초경	TE	Triple Heater Meridian
11	족소양담경	GB	Gall Bladder Meridian
12	족궐음간경	LR	Liver Meridian

11. Face Meridian Chart

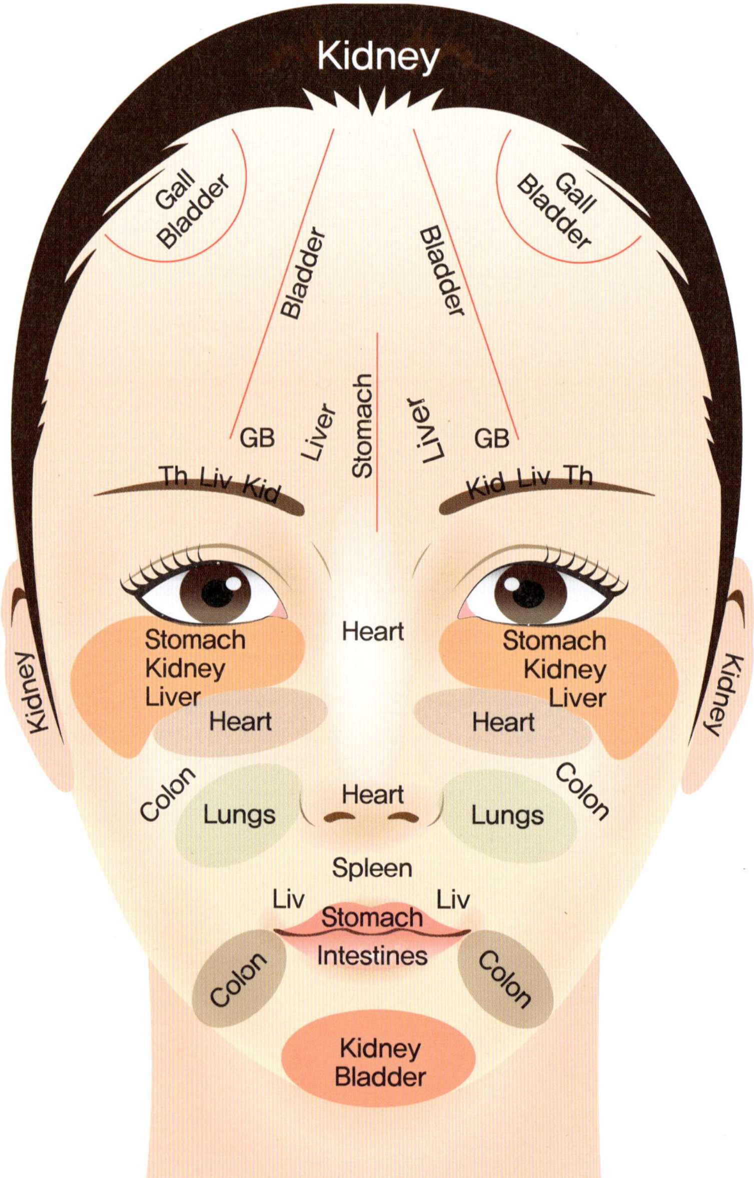

12. Lymph Drainage

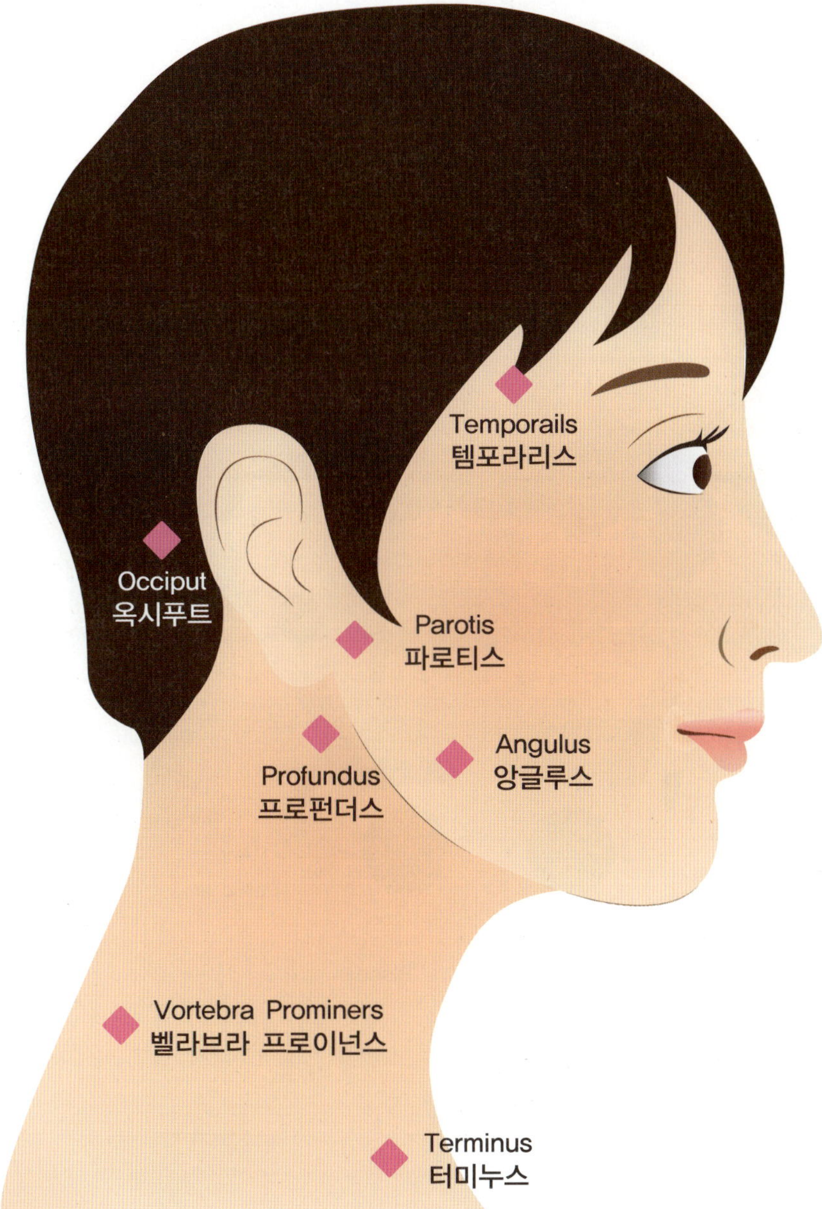

13. Foot Reflexology Points

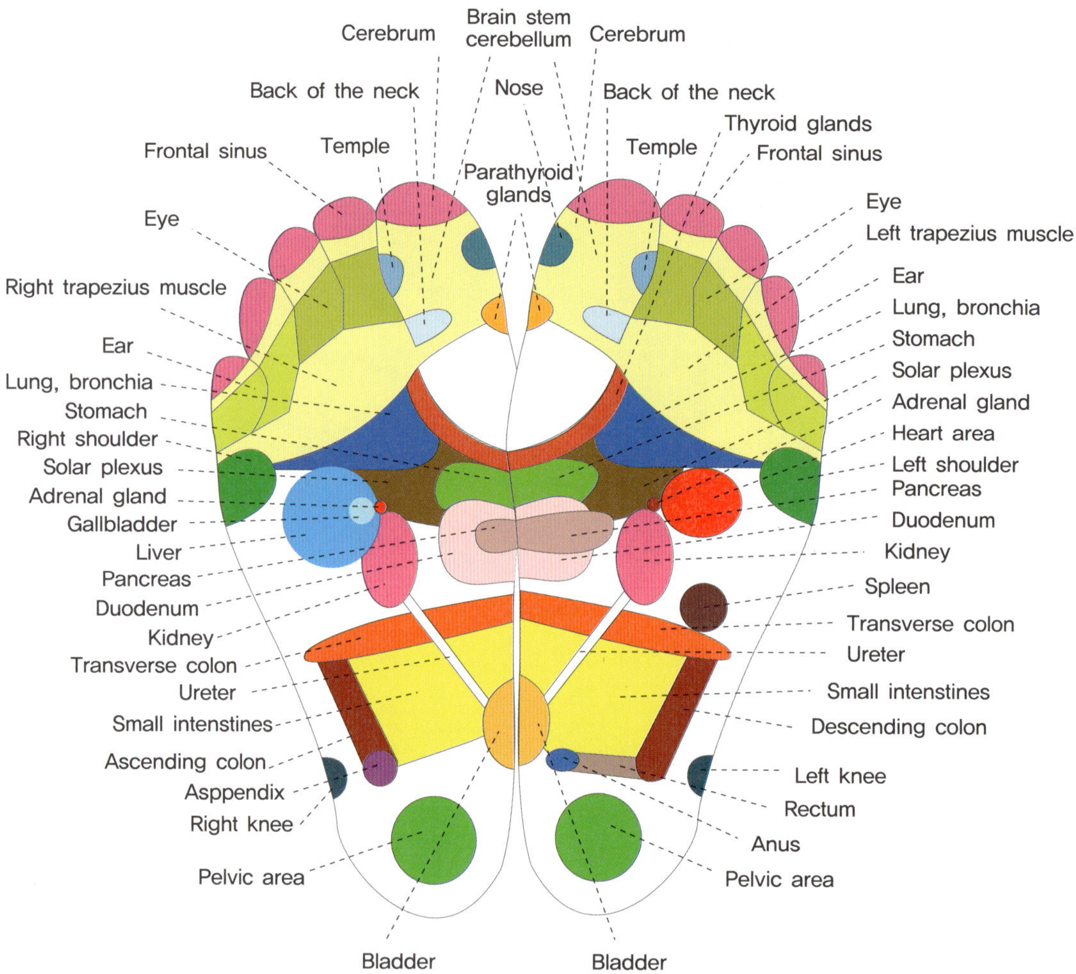

14. Materials & Tools for makeup

<table>
<tr><td>
</td><td></td><td>
</td><td></td></tr>
<tr><td>Foundation</td><td>Concealer</td><td>Powder</td><td>Lip-Stick</td></tr>
<tr><td></td><td></td><td>
</td><td></td></tr>
<tr><td>Lip Palette</td><td>Eye shadow Palette</td><td>Eyeliner</td><td>Eyebrow Pencil</td></tr>
<tr><td></td><td>
</td><td></td><td>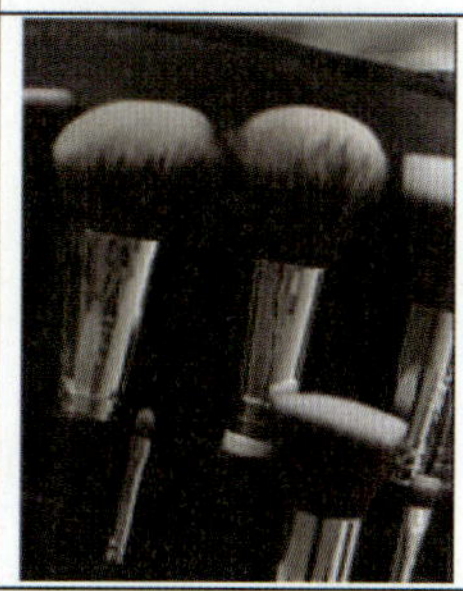</td></tr>
<tr><td>Mascara</td><td>Lip & Eye Makeup Remover</td><td>Sponge</td><td>Makeup brush</td></tr>
<tr><td colspan="2"></td><td colspan="2"></td></tr>
<tr><td colspan="2">Brush set</td><td colspan="2">Make-up Box</td></tr>
</table>

15. Materials & Tools for Nail care

	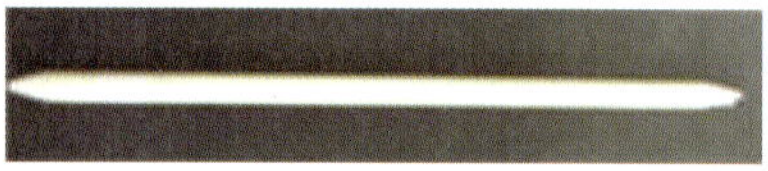		
Cuticle Pusher	Orange Wood Stick		Emery Board
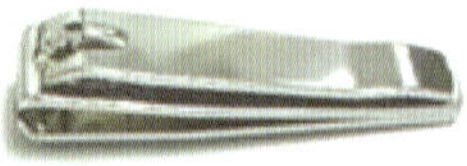	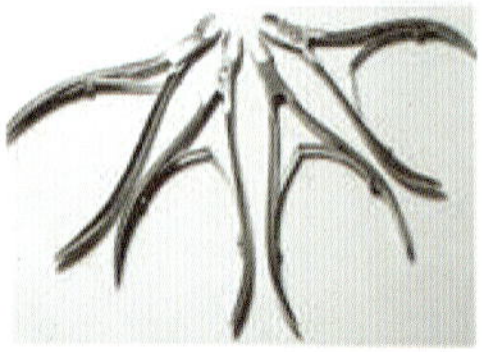		
Clipper	Cuticle Nipper	Nail Brush	Buffer

Antiseptic	Gel Glue	Polish Remover	Cuticle Oil

Nail Lacquer	Nail Tip	Acrylic Powder	Nail Dryer

16. Color Name the Backbone of the ISCC-NBS System

10 Hue and three neutral categories(13 basic color categories)

pink (Pk)	olive (Ol)	white (Wh)
red (R)	yellow green (YG)	gray (Gy)
orange (O)	green (G)	black (Bk)
brown (Br)	blue (B)	
yellow (Y)	purple (P)	

16 Intermediate categories

reddish orange (rO)	purplish blue (pB)	reddish brown (rBr)
orange yellow (OY)	violet (V)	yellowish brown (yBr)
greenish yellow (gY)	reddish purple (rP)	olive brown (OlBr)
yellowish green (yG)	purplish red (pR)	olive green (OlGr)
bluish green (bG)	purplish pink (pPk)	
greenish blue (gB)	yellowish pink (yPk)	

Part Ⅱ

Telephone Reservation & Business Hours Inquiry

Business English for Beauticians

CHAPTER 01

Hair & Scalp Care

1. Business Hours Inquiry

Manager : Thanks for calling Kelly' s hair salon.

Customer : Hello. I' d like to know what your business hours are.

Manager : The business hours are 9 am to 8 pm Monday through Saturday and we are closed on Sundays.

Customer : Are there any special event?

Manager : Currently we are running a special event on perms to celebrate our seventh anniversary. If you have your hair permed between 9 am and 12 am this week, you can save 10%.

Customer : OK. I' d like to make an appointment for a perm at 9 am tomorrow.

Manager : Sure. Do you have a hair stylist?

Customer : No, I have not.

Manager : No problem. I' ll set you up with Michelle. She' ll take good care of you. Let me confirm the time and date again. Tomorrow for a perm at 9 am. Is that right?

Customer : Right, thanks.

Manager : OK. See you tomorrow. Bye.

Key words

anniversary : 기념일
celebrate : 축하, 경축하다
confirm the time and date : 시간과 날짜를 정하다
set up : 배정하다, 설정하다
take good care of : 잘 보살피다 = look after

제1장 모발 및 두피 관리

1. 영업시간 문의

매니저 : 감사합니다. 켈리 미용실입니다.

고객 : 여보세요. 영업시간이 언제인지 알 수 있을까요?

매니저 : 저희 영업시간은 월요일부터 토요일까지는 오전 9시부터 오후 8시까지이고, 매주 일요일은 휴무입니다.

고객 : 특별 이벤트가 있나요?

매니저 : 네. 현재 저희 미용실은 개점 7주년을 축하하기 위해 특별 사은행사로 펌을 할인가로 해드리고 있습니다. 이번 주 오전 9시에서 12시 사이에 펌을 하시면 가격의 10%를 할인해 드립니다.

고객 : 그렇군요. 내일 오전 9시에 펌 예약을 했으면 하는데요.

매니저 : 네. 담당 스타일리스트가 있으십니까?

고객 : 아니요, 없는데요.

매니저 : 걱정마세요. 제가 미셸 선생님으로 예약해 드리겠습니다. 미셸 선생님이 잘 해드릴 겁니다. 날짜와 시간을 다시 확인해보겠습니다. 내일 아침 9시에 펌예약 맞죠?

고객 : 네, 맞습니다. 고마워요.

매니저 : 그럼 내일 뵐게요.

2. Price Inquiry

Manager : Thanks for calling Kelly' s hair salon.

Customer : Hello. Can I ask how much a perm is?

Manager : What type of perm?

Customer : Setting Perm.

Manager : Could you tell me how long your hair is?

Customer : It touches my shoulder. Is the price different depending on hair length?

Manager : Yes. The price depends on your hair length and the quality of perm solution. Your hair is medium length. If you use a regular perm solution on medium length hair, it is 80,000 Won. But if you want to use the best quality perm solution, it is an additional 10,000 Won. Using a little lower quality, the price will be lower too.

Customer : Does that price include a hair cut?

Manager : No. You' d pay for a hair cut separately.

Customer : How much will that cost?

Manager : A hair cut for a perm is 10,000 Won, a lower price than that of a regular hair cut. In addition, we offer a 10% discount for students and senior citizens on coloring.

Customer : Ok. Thank you for your help. Bye!

Manager : Thanks for your calling, Bye

Key words

additional : 추가적인
in addition : 게다가
depending on : ...에 따라
medium length : 중간 길이
quality : 품질, 양질
perm solution : 펌 액
senior citizen : 노인

2. 가격 문의

매니저 : 감사합니다, 켈리 미용실입니다.

고객 : 안녕하세요? 펌 가격이 얼마인지 알 수 있을까요?

매니저 : 어떤 펌을 하실 건가요?

고객 : 셋팅 펌이요.

매니저 : 실례지만 고객님의 모발 길이가 어느 정도 되나요?

고객 : 어깨에 닿을 정도에요. 모발 길이에 따라 가격도 달라지나요?

매니저 : 네. 펌 가격은 모발 길이와 펌 용액의 품질에 따라 다릅니다. 고객님의 머리는 중간 길이이죠. 중간 길이에 일반 펌 용액을 사용하면 가격은 8만 원입니다. 하지만 가장 좋은 품질의 펌 용액을 사용하시면 일만 원 추가됩니다. 물론 좀 더 낮은 등급을 사용하시면 가격도 더 낮아집니다.

고객 : 그럼 그 가격은 커트 비용도 포함된 것 인가요?

매니저 : 아니요. 커트 가격은 따로 지불하셔야 합니다.

고객 : 커트 가격은 얼마인데요?

매니저 : 펌을 할 때 커트를 하시면 만원인데, 일반 커트만 할 때보다 더 싼 가격입니다. 게다가 학생이나 노인 분들이 염색을 하는 경우는 10퍼센트 할인 혜택을 드리고 있습니다.

고객 : 네, 알려주셔서 고맙습니다. 안녕히 계세요.

매니저 : 전화 주셔서 감사합니다. 안녕히 계세요.

3. Reservation for coloring

Manager : Thanks for calling Kelly' s hair salon.

Customer : Hello. This is Mrs. Jolly. I' d like to make an appointment for Thursday?

Manager : Sure. Who' s your stylist?

Customer : Ms. Chanel.

Manager : What time works for you?

Customer : How about 1 pm?

Manager : Hold on a second, let me check her availability. I' m really sorry. Ms. Chanel is booked all afternoon that day.

Customer : Then, what time is good?

Manager : She has some time in the morning at 10 o' clock and an opening in the evening at 7 o'clock.

Customer : OK. 10 o' clock works for me.

Manager : OK. What are you looking to have done?

Customer : Just a cut and color.

Manager : OK. I' ll put you down at 10 o' clock in the morning on Thursday for a cut and color.

Customer : OK.

Manager : All right, you are all set. Is there anything else that I can help you?

Customer :Hmm.... Can you tell me how to get there?

Manager : Sure. Where are you coming from?

Key words

make an appointment = set an appointment : (시간) 예약하다 cf. make a reservation : (장소) 미리 잡아두다, 예약하다
am = ante meridiem : 오전
available : 이용할 수 있는, 쓸모 있는, 유효한
availability : 이용도, 이용 가능성
coloring : 염색하다 = pigmentation, dyeing
get off : 내리다
hold on a second 잠깐만 기다리세요 (전화상에서)
be located : ...에 위치하다
opening time : 영업 시작 시간
pm = post meridiem : 오후
prefer : ....을 더 좋아하다
put down : 적어 놓다
take the subway : 지하철을 타다
take the taxi : 택시를 타다
take exit number 7 : 7번 출구로 나오다

3. 염색 예약

매니저 : 켈리 뷰티살롱에 전화 주셔서 감사합니다. 무엇을 도와드릴까요?

고객 : 여보세요? 저는 졸리부인 인데요. 목요일로 예약을 좀 하고 싶은데요.

매니저 : 그러세요. 담당 스타일리스트가 누구세요?

고객 : 샤넬선생님인데요.

매니저 : 언제가 좋으십니까?

고객 : 오후 1시쯤 어떠세요?

매니저 : 잠깐 기다리세요. 가능한지 체크해 보겠습니다. 정말 죄송합니다만 샤넬 선생님은 목요일 오후 내내 예약이 꽉 찼는데요.

고객 : 그럼 언제가 좋을까요?

매니저 : 아침 10시하고 저녁 7시에 샤넬 선생님의 예약이 비어 있습니다.

고객 : 10시도 괜찮아요.

매니저 : 알겠습니다. 무엇을 하실 건가요?

고객 : 그냥 좀 자르고 염색 좀 하려고요.

매니저 : 네. 목요일 아침 10시 커트와 염색을 하시는 걸로 예약해 드리겠습니다.

고객 : 네

매니저 : 네, 전부 끝났습니다. 다른 더 도와드릴 사항은 없습니까?

고객 : 음... 거기 가려면 어떻게 해야 하나요?

매니저 : 어디에서 오실 건데요?

Customer : I'll be taking the subway from Yongsan.

Manager : We are located near Kangnam station. Take number 2 and get off at Kangnam station and then you must take exit number 7 and walk to your left. You'll see the pink beauty salon building.

Customer : Oh, I see. Thank you for your help.

Manager : You're welcome. Then we will see you on Thursday. Bye.

Customer : Bye.

4. First Visit

Manager : Hello! Welcome to Kelly's Hair Salon. What can we do for you?

Customer : I'd like to get a hair cut.

Manager : OK. I am the manager Minjeong Kim. It's a pleasure to meet you.

Customer : I'm Kate. Nice to meet you, too.

[A 경우]

Manager : This is Sam. She is a hair cut and perm specialist.

Sam : Pleasure to meet you. You wanted a hair cut today. Tell me how you'd like it done.

Customer : I want a good trim because I have a lot of new growth.

Sam : Sure.

Key words

customer : (가게의) 손님, 고객
introduce : 소개하다
manager : 지배인, 관리자 (director)
at home = comfortable
specialist : 전문가 (= expert)
Can I interest you in~? 어떠십니까?

고객 : 용산에서 지하철을 타고 갈 거예요.

매니저 : 저희 살롱은 강남역 근처에 있습니다. 지하철 2호선을 타셔서 강남역에서 내리세요. 그리고는 7번 출구로 나와 좌측방향으로 걸어오시면 됩니다. 분홍색 뷰티 살롱 건물이 바로 보일 겁니다.

고객 : 오, 알겠습니다. 도와주셔서 감사합니다.

매니저 : 별 말씀을요. 그럼 목요일 아침에 뵙겠습니다.

고객 : 수고하세요.

4. 첫 방문

매니저 : 어서 오세요. 무엇을 도와 드릴까요?

고객 : 네, 머리 커트를 하려고 합니다.

매니저 : 그러세요. 저는 이 미용실의 매니저 김민정이라고 합니다. 반갑습니다.

고객 : 저는 케이트예요, 반갑습니다.

[A 경우]

매니저 : 이 분은 쌤입니다. 커트와 펌 전문가입니다.

쌤 : 반갑습니다. 오늘은 커트를 하신다고요? 어떤 머리형을 원하시나요?

고객 : 머리가 많이 길어서 그러는데 가지런히 다듬어주세요.

쌤 : 알겠습니다.

[B 경우]

Manager : Kate, we have Edward Kim, Chanel Lee and Vivian Cho as hair cut specialists, and Jack Hong as a hair perm specialist.

Customer : Emm.. My friend introduced to me Chanel Lee.

Manager : She' ll be with you in 5 minutes. Would you mind waiting?

Customer : Of course.

Manager : Can I interest you in something to drink while you wait? We have coffee, green tea, and juice. Which do you prefer?

Customer : Coffee sounds good, thanks.

Manager : Sure. Here are magazines. Make yourself at home.

Customer : Thank you.

5. Waiting Time

Manager : Miss Kim, I' m sorry. Your hair designer is still finishing up with her/his previous Customer. Please wait here a moment.

Customer : Okay.

Manager : Thank you. We can store your valuables. Would you like to leave your things with us?

Customer : Yes, here you are.

Manager : Let me take your belongings, Miss Kim. Here is your locker key. Your locker number is 7.

[B 경우]

매니저 : 에드워드 김, 샤넬 리, 비비안 조 선생님이 커트 전문이고, 젝 홍선생님은 펌 전문입니다. 혹시 원하시는 디자이너 선생님이 있습니까?

고객 : 음... 제 친구가 샤넬 리 선생님을 소개해 주셨습니다.

매니저 : 샤넬 리 선생님은 5분 정도 후에 오실 수 있습니다. 잠시만 기다려 주시겠습니까?

고객 : 그럴게요.

매니저 : 기다리는 동안 마실 것 좀 드릴까요? 커피와 녹차, 주스가 있는데 뭐가 좋으세요?

고객 : 커피요.

매니저 : 네, 알겠습니다. 여기에 잡지가 있으니 보면서 편안히 기다려주세요.

고객 : 고맙습니다.

5. 대기 시간

매니저 : 고객님, 죄송합니다. 담당 디자이너가 이전 고객 시술을 마무리하는 중입니다. 여기서 잠시만 기다려 주시겠습니까?

고객 : 그러죠.

매니저 : 고맙습니다. 저희가 고객님의 귀중품을 보관해 드릴 수 있습니다. 소지품을 저희에게 맡기시겠습니까?

고객 : 네, 그러겠습니다.

매니저 : 고객님, 제가 보관해 드리겠습니다. 여기 보관함 열쇠가 있습니다. 보관함 번호는 7번입니다.

Key words

1. cosmetologist
   - **a hair dresser**
     = a beautician,
     a beauty operator,
     a beautician
   - **a cosmetician**
     - skin care, makeup work
     - a makeup artist
     - esthetician
       = a skin care specialist,
       a skin care expert
   - **a manicurist**
     - a nail care artist/expert
   - **an electrologist**
   - **a sharone manager**
   - **a manufacturer's technician**
     - technical consultant
   - **a platform artist**

2. 기타
   - **a masseur**
   - **manager**
   - **patron**
   - **customer**
   - **client/ clientele**
   - **belongings/ personals**
   - **valuables**
   - **essentials**

Key words

finish up : 끝내다
pleasure : 기쁨, 만족
previous : 이전의, 먼저
personal belongings : 개인 소지품
store : 보관, 저장하다
valuables : 귀중품

Customer : Thank you very much.
Manager : Would you like something to drink?
Customer : Sure. What do you have?
Manager : We have juice, coffee and green tea.
Customer : I'll have an iced green tea, please.
Manager : Okay, I' ll be right back with your drink. Here are some magazines to read while you wait.
Customer : Thank you so much.
Manager : It's my pleasure.

6. Evaluation and Payment

Manager : How was everything today?
Customer : Everything was great. Thanks.
Manager : I am glad to hear that. Is there anything else we can do for you today?
Customer : No. That' s all.
Manager : OK. May I have your locker key, please?
Customer : Here you are.
Manager : Here are your belongings, Jinny. Please make sure you have everything.
Customer : It's all here. What do I owe you?
Manager : You had a regular hair trim and hair care treatment today. Is that right? Your total is fifty thousand won. Would you like to pay with cash or card?

고객 : 감사합니다.

매니저 : 음료를 드시겠습니까?

고객 : 네, 무슨 종류의 음료가 있나요?

매니저 : 주스, 커피 그리고 녹차가 있습니다.

고객 : 전 냉 녹차로 하겠습니다.

매니저 : 네, 음료 바로 가지고 오겠습니다. 여기 기다리시는 동안 보실 잡지입니다.

고객 : 감사합니다.

매니저 : 별 말씀을요.

6. 계산 및 지불

매니저 : 오늘 모든 시술은 어떠셨습니까?

고객 : 모든 것이 좋았습니다. 고마워요.

매니저 : 그렇게 말씀해 주시니 기쁩니다. 더 필요한 것은 없으십니까?

고객 : 네, 없어요.

매니저 : 보관함 열쇠를 주시겠습니까?

고객 : 여기 있습니다.

매니저 : 여기 고객님의 물건이 있습니다. 모든 물건이 그대로인지 확인해 보시겠어요?

고객 : 다 그대로 있네요. 얼마죠?

매니저 : 고객님은 오늘 커트와 머리 관리를 하셨습니다. 맞으시죠? 모두 5만 원입니다. 계산은 현금과 신용카드 중 어느 것으로 하시겠습니까?

Customer : Is Visa card OK?

Manager : Of course. Would you like to pay in installments?

Customer : I' d like to pay all at once.

Manager : Ok. Check your bill and sign here, please.

Customer : Thank you.

Manager : Here' s my card. If you want to make an appointment, please give me a call anytime.

Customer : Thank you.

Manager : You' re welcome. I hope to see you next week, Jinny. Have a great day.

Customer : Bye.

Key words

bill : 계산서, 청구서
credit : 신용의
payment : 지불, 계산
prefer : 선호하다
pay in installments : 할부로 하다
sign : 서명하다. 간판
signature : 편지 말미 자필 서명
cf. autograph- (연예인) 사인
What do I owe you?
당신한테 줄 돈이 얼마죠?
How much is it all together?
모두 얼마죠?
What is the price?
How much is it?
비용이 얼마죠?
Can I pay half cash and half card?
제가 절반은 현찰 그리고 절반은 카드로 지불해도 될까요?
make one payment : 지불하다
= pay all at once : 한번에 불하다
= pay in full : 전액을 지불하다.

고객 : 비자카드도 되나요?

매니저 : 물론이죠. 할부로 하시겠습니까?

고객 : 전 일시불로 하고 싶습니다.

매니저 : 네, 알겠습니다. 계산서 확인하시고, 여기에 서명 부탁드립니다.

고객 : 감사합니다.

매니저 : 이것은 제 명함입니다. 예약을 원하시면 언제든지 제게 전화주세요.

고객 : 감사합니다.

매니저 : 별 말씀을요. 다음 주에 뵙길 바랍니다, 지니님! 좋은 하루 보내세요.

고객 : 안녕히 계세요.

Let's Talk

◎ 처음 만났을 때 인사하는 표현을 익히고, 주어진 말을 활용하여 대화해 봅시다.

A : Hello, welcome to Sharone' s hair Salon. I am the manager Julia Kim.

B : Hi, I am Fanny.

A : It' s a pleasure to meet you.

B : Nice to meet you, too.

Reference

만나서 반갑습니다.

It' s nice to meet you.

I' m happy/glad to meet you.

I' m pleased to meet you.

I' m honored to meet you.

I' m so delightful to meet you.

◎ 소개할 때 필요한 표현을 익히고 주어진 말을 활용하여 대화해 봅시다.

Manager : This is Jinny. She is a haircut and perm wave specialist.

Hair Designer : Pleasure to meet you.

Reference

I' ll be your hair beautician today. My name is Candy.

I' ll be doing your hair today.

Candy will be your beautician today.

◎ 어떤 디자이너를 원하십니까?

① 담당 디자이너가 있으신가요?

Who is your beautician?

Do you have a hair designer?

② 담당 디자이너의 이름이 무엇인가요?

What' s the name of your beautician?

③ 마리나입니다.

Her name is Marina.

④ 6번을 찾습니다.

It' s number 6.

⑤ 혹시 담당 디자이너의 명함을 가져오셨나요?

Do you have your beautician' s card?

⑥ 여기 있어요.

Here you are.

⑦ 염색을 잘하는 디자이너로 소개해주세요.

Someone who specializes in colouring.

⑧ 이 분이 파마를 제일 잘 합니다.

This beautician specializes in perming.

⑨ 리처드가 커트를 제일 잘합니다.

This is Richard, he specializes in cutting.

◎ 성함이 어떻게 되세요?

Can I take your name please?

May I ask your name?

Can I ask your name?

◎ 제가 고객님의 머리를 해드릴 담당 헤어 디자이너 캔디입니다.

I' ll be your hair beautician today, My name is Candy.

I' ll be doing your hair today.

Candy will be your beautician from now on.

① 여기는 저를 도와줄 모니카입니다.

This is my apprentice, her name is Monica.

She/He will be helping me today.

② 저는 고객님 담당 미용사를 도와줄 모니카입니다.

I am your beautician' s assistant. My name is Monica.

I am your beautician' s apprentice.

M : Is this your first time here?

C : Yes, it is. I would like a new style.

M : Do you want any particular beautician?

C : I' d like an experienced beautician, please.

M : Okay. Let me introduce your beautician to you, this is Maria.

◎ 예약할 때

C : Is this Sharone' s Beauty Salon?

M : Yes, it is.

C : Can I make an appointment for later today?

M : Sure, we have an opening between 3 and 4.

C : I' d like to make it at 3: 30, then.

M : All right, your name, please?

C : This is Mrs. Kim.

M : Okay, Mrs. Kim, see you at 3: 30.

◎ 그 밖의 회화

① 오늘 오후 5시로 예약을 하고 싶습니다.

I' d like to make an appointment for 5 P.M. today.

② 이 호텔 안에 미용실(이발소)이 있습니까?

Is there a beauty shop (barbershop) in this hotel?

③ 오늘 오후에 예약을 할 수 있을까요?

Can I make an appointment for this afternoon?

④ 예약을 해야 하나요?

Do I need a reservation?

⑤ 예약을 하지 않고도 머리할 수 있나요?

Do you have time for a walk-in?

⑥ 파마는 예약 없이 해드릴 수 가 없습니다.

Hair perm wave service is not offered on a 'walk-in' basis without reservations.

⑦ 미셸은 오늘은 예약이 모두 찼는데요.

Michelle is all booked today.

M : Hello, did you have an appointment with us?
C : No, but do you have time for a walk-in?
M : I' m afraid I can' t. We are all booked up right now.
If you don' t mind, how about after one hour or so?
C : Okay.

⑧ 제가 좀 일찍 왔어요.

I' m a bit early.

S : Hello, do you have an appointment?

C : Yes, but I' m a bit early. I have a 1 o' clock appointment.

⑨ 오늘 처음 오셨습니까?

Is this your first time here?

Is this your first visit to us?

Are you a new client?

⑩ 아니오, 예전에 왔었어요.

No, I' ve been before.

⑪ 아니오, 두 번째에요.

No, this is my second time.

⑫ 혹시 원하는 디자이너가 있으세요?

Have any of our beauticians been recommended to you?

Do you want any particular beautician?

⑬ 캔디가 잘한다고 들었어요.

I hear that Candy is excellent.

Is Candy available? 주의) Are you available?

⑭ 경험이 많은 분으로 해주세요.

I' d like an experienced beautician.

⑮ 누구든 잘하면 상관없어요.

Anyone good will be fine.

⑯ 빨리 할 수 있는 사람이면 좋겠어요.

I' ll see anyone who' s available.

⑰ 당신이 추천해 주시겠어요?

Who do you suggest?

Who would you suggest?

⑱ 안내해드릴게요. 이쪽으로 와주세요.

(If you' d like to) follow me.

Please come this way.

⑲ 잠시만 앉아서 기다려주세요.

Please take a seat.

Please sit down.

⑳ 담당디자이너가 곧 올 겁니다.

Your beautician will be with you in a moment.

A beautician will be with you soon.

Exercise. Fill in the blanks!

1. 성함이 어떻게 되세요?
   Can I (   ) your name please?
2. 제가 고객님의 머리를 해드릴 담당 헤어 디자이너 캔디입니다.
   I' ll be your (   ) (   ) today, My name is Candy.
3. 담당 디자이너가 있으신가요?
   Do you have ( ) (   ) (   )?
4. 담당 디자이너의 이름이 무엇인가요?
   What' s ( ) (   ) of your beautician?
5. 빨리 할 수 있는 사람이면 좋겠어요.
   I' ll see anyone who' s (   ).
6. 경험이 많은 분으로 해주세요.
   I' d like an (   ) beautician.
7. 혹시 담당 디자이너의 명함을 가져오셨나요?
   Do you have your beautician' s (   )?
8. 여기 있어요.
   (   ) it is.
9. 염색/퍼머를 잘하는 디자이너로 소개해주세요.
   Someone who (   ) ( ) colouring /perming.
10. 누구든 잘하면 상관없어요
    Anyone (   ) will be (   ).

Answer

1. take
2. hair, stylist
3. a, hair, designer
4. the, name
5. available
6. experienced
7. card
8. Here
9. specializes, in
10. good, fine

11. 여기는 저를 도와줄 미니입니다.

This is my (   ), her name is Mini.

12. 저는 고객님 담당 미용사를 도와줄 미니입니다.

I am your beautician' s (   ). My name is Mini.

13. 오늘 오후 5시로 예약을 하고 싶습니다.

I' d like to (   ) (   ) (   ) for 5 P.M. today.

14. 이 호텔 안에 미용실이 있습니까?

Is there ( ) (   ) (   ) in this hotel?

15. 당신이 추천해주시겠어요?

Who would you (   )?

Make scrambled words into a correct sentence!

16. how / your / done / Tell / you' d / me / like / hair

17. you / minutes / She'll / with / in / be / 5

18. new / growth / I / new / a / of / have / lot

19. at / yourself / Make / home

20. in / drink / Can / to / waiting / while / you / I / something / you / are / interest?

Answer

11. apprentice
12. assistant
13. make, an, appointment
14. a, beauty, salon
15. suggest
16. Tell me how you' d like your hair done.
17. She' ll be with you in 5 minutes.
18. I have a lot of new growth.
19. Make yourself at home.
20. Can I interest you in something to drink while you are waiting?

CHAPTER 02

Skin Care

1. Price Inquiry

Manager : Esthetic Salon Healing. This is Marion. May I help you?

Customer : Hello, I' d like to ask what kinds of facial skin care do you offer.

Manager : We offer 3 different facial skin care programs. Normal program divide in the 60 minute and 80 minute treatment. The 60 minute facial skin care consists of skin analysis, exfoliation of death cell, swedish massage and anti-wrinkle treatment. It costs 140,000 Won. The 80 minute facial care include detox, anti-aging treatment, korean massage plus a scalp and foot massage. It costs 200,000 Won. The last special programs consist of aroma therapy and swedish massage for sensitive skin type and combination skin type. It costs 200,000 Won.

Customer : Thank you. But I have one more question. Can you explain detail about the special program?

Manager : Of course. We use natural essential oil and carrier oil for the aromatherapy, and safety guaranteed organic cosmetics for manual massage.

Key words

divide : 나누다, 분열하다.
exfoliation : 박리, 벗겨짐
guaranteed : 확보된, 보증된
healing : 치유, 치료
safety : 안전, 안전성
sensitive : 예민한, 민감한

제2장 피부관리

1. 가격 문의

매니저 : 힐링 피부관리실 마리온입니다. 무엇을 도와 드릴까요?

고객 : 안녕하세요. 어떤 종류의 얼굴관리가 있는지 알고 싶은데요.

매니저 : 저희는 3가지 얼굴관리 프로그램이 있습니다. 일반적인 프로그램은 2가지가 있습니다. 60분 관리는 피부분석, 각질제거, 스웨디시 마사지와 주름개선 관리로 구성되어 있습니다. 가격은 14만원입니다. 80분 관리는 해독, 항노화 관리, 한국형 마사지에 두피와 발마사지가 포함되어 있습니다. 가격은 20만원입니다. 특별마사지는 아로마요법과 민감성 피부와 복합성 피부를 위한 스웨디시 마사지로 구성되어 있습니다. 가격은 20만원입니다.

고객 : 그런데 질문 하나가 더 있는데요. 특별 관리에 대해서 좀 더 상세하게 설명해 주시겠어요?

매니저 : 네 물론이죠. 아로마요법에는 천연 에센셜오일과 캐리어 오일이 사용되고, 매뉴얼 마사지에는 안전성이 확보된 유기농 화장품이 사용됩니다.

Customer : I have an interest in special program. Can you make an appointment for me the day after morning?

Manager : Today is Monday. Wait a moment. How is your reservation time at 10 am on wednesday?

Customer : Okay, it' s good. Thank you for your help. Bye!

Manager : Thanks for your calling. See you on wednesday, Bye.

2. Reservation Ⅰ

Manager : Beauty Salon. This is Cathy. May I help you?

Customer : Hello, I' d like to make an appointment for a facial skin care tomorrow.

Manager : Of course, what time would you like to come in?

Customer : Sometime tomorrow morning.

Manager : We are almost all booked up tomorrow. Please give me a moment. Let me check our availability. We have an opening at 10:30. How about 10:30?

Customer : 10:30? That sounds great. Also, what kind of treatments do you offer?

Manager : We have two options. The 60 minute facial is a basic treatment that costs 90,000 Won. This treatment consists of two masks and a facial massage. The 90 minute facial consists of the basic treatment plus a scalp, hand and arm, and foot massage. It costs 180,000 Won.

Key words

book : (명) 책, (동) 예약하다
be booked up : 예약이 꽉 차다
consist of : ...로 구성되다
facial : 얼굴 마사지
have a great day! : 좋은 하루되시길!
option : 선택사항

고객 : 저는 특별관리에 관심이 있습니다. 모레 저를 위해서 약속을 잡아 주시겠어요?

매니저 : 오늘은 월요일이고요. 잠깐만 기다려 주세요. 예약시간이 수요일 오전 10시가 어떠신지요?

고객 : 네 좋습니다. 도와주셔서 감사합니다. 안녕히 계세요.

매니저 : 전화 주셔서 감사합니다. 수요일에 뵙겠습니다. 안녕히 계세요.

2. 예약 I

Check point

미용산업에서 고객과의 첫 만남은 전화예약으로 이루어지는 것이 대부분이며, 이는 살롱의 첫 이미지를 좌우한다 해도 과언이 아니다. 간단한 전화 내용을 토대로 이를 응용하여 살롱의 실제 상황에서 전화 서비스를 효과적으로 수행할 수 있도록 한다.

매니저 : 뷰티 살롱의 캐시입니다. 무엇을 도와드릴까요.

고객 : 안녕하세요? 내일 피부 관리 예약을 하고 싶습니다.

매니저 : 네, 언제로 예약하고 싶으세요?

고객 : 내일 아침에요.

매니저 : 내일 예약은 거의 다 찼는데요, 잠시만 기다려 주십시오. 스케줄을 살펴보겠습니다. 10시 30분에 빈 시간이 있는데 10시 30분 어떠세요?

고객 : 10시 30분이요? 좋아요. 구체적으로 제공하는 서비스 내용이 무엇인가요?

매니저 : 60분 피부 관리는 9만원으로 기본적인 관리입니다. 이 관리는 두 가지 종류의 마스크와 얼굴 마사지로 이루어집니다. 90분에 해당하는 피부 관리는 기본적인 관리에 두피, 손과 팔, 그리고 발 마사지가 추가로 이루어집니다. 가격은 18만원입니다.

Customer : I' ll have the 90 minute facial.

Manager : Certainly! So we will see you tomorrow at 10 :30. If you' re late, we' ll start without you. Thank you for calling and see you tomorrow. Have a great day! Bye!

3. Reservation II

Manager : This is Trendy Salon. My name is Kate. May I help you?

Customer : Hello. I want to make a reservation for a facial tomorrow.

Manager : Sure, what time works for you?

Customer : Is tomorrow morning OK?

Manager : Sure, hang on a moment. Let me check our availability for tomorrow. You' re in luck. We' re all booked in the afternoon but we do have time in the morning at 10:30. Does that suit you?

Customer : It' s a little early but it' ll do. Also, what are the different massages you offer?

Manager : We offer a basic facial, a Korean style facial, a clarifying facial and an anti-aging facial. Everyone' s skin is unique and requires a different level of treatment and care. Because of this, the price will vary depending on your skin type and needs. The most basic skin care starts at around 50,000 won. Which one do you have in mind?

Key words

clarifying : 명확하게 하다, 투명하게 만들다
decide : 결정하다
grain : 곡물, = cereal
hang on : 잠시 기다리다 (전화를 끊지 않고 두다)
include : 포함하다

고객: 90분 피부 관리 서비스로 할게요.

매니저: 네. 그러면 내일 10시 30분에 뵐게요. 시간 늦지 마시기 바랍니다. 전화 주셔서 감사드립니다. 그럼 내일 뵙겠습니다. 좋은 하루 보내십시오.

3. 예약 II

매니저: 트렌디 살롱입니다. 저는 케이트입니다. 무엇을 도와드릴까요?

고객: 안녕하세요. 내일 피부 관리 예약을 하고 싶습니다.

매니저: 네, 언제가 좋으세요?

고객: 내일 아침에 가능할까요?

매니저: 네, 잠시만 기다려 주세요. 내일 가능한지 스케줄을 살펴보겠습니다. 네, 다행히도 내일 오후 예약은 다 되어있습니다만, 오전 10시 반에는 시간이 비었네요. 어떠십니까?

고객: 좀 이르기는 하지만, 그렇게 할게요. 그리고 어떤 서비스들을 제공하시나요?

매니저: 기본 관리인 매뉴얼마사지가 있고, 한국형 마사지, 피부를 맑게 해주는 마사지 그리고 노화 방지를 위한 마사지가 있습니다. 사람들마다 피부가 다르고 여러 가지 관리를 필요로 합니다. 그 때문에 가격은 각 피부 타입이나 필요사항에 따라 다릅니다. 가장 일반적인 피부 관리는 5만 원 선부터 있습니다. 어떤 서비스를 염두에 두고 계시나요?

Customer : I am sorry but I don' t know. What are the specifics of each skin care?

Manager : The basic facial massage costs 50,000 won and consists of a basic face massage and two masks, a grain and a honey mask. You can think of it as a basic skin management lasting 60 minutes. If you want a stronger facial, you can get a Korean style. However, since the Korean style facial and the clarifying facial and the anti-aging facial include additional services, the cost will be a little higher. Which facial do you feel you need?

Customer : I think I' ll decide on the specific massage after our consultation tomorrow. Is that okay?

Manager : Of course. Your time is 10:30. After your consultation, I' m sure you' ll know which one will meet your needs. Please give us a call if you' re running late.

고객 : 미안하지만 잘 모르겠어요. 각각의 서비스는 특징이 뭔가요?

매니저 : 가장 일반적인 관리 비용은 5만원으로, 기본 얼굴 마사지와 두 가지 마스크(곡물마스크와 벌꿀 팩)로 이루어집니다. 60분 정도의 기본적인 피부 관리라고 생각하시면 됩니다. 만약 조금 더 집중적인 마사지를 받고 싶으시다면 한국형 마사지를 받으시면 됩니다. 하지만 특수마사지인 한국형 마사지와 피부를 맑게 해주는 마사지와 노화 방지 마사지는 추가비용을 내야하기 때문에 비용은 좀 더 추가 될 거예요. 어떤 피부 관리를 받고 싶으세요?

고객 : 내일 방문해 상담 후 특수마사지로 결정할게요. 가능하죠?

매니저 : 알겠습니다. 고객님의 내일 예약 시간은 오전 10시 30분입니다. 상담 후에 어떤 마사지가 고객님 피부와 맞을지 알게 될 거예요. 만약 늦어지시면 전화주세요.

4. SPA Reservation

Manager : Good afternoon. Sharone SPA. May I help you?

Customer : Hello. I am Sunny. I' d like to know what kinds of facilities I can use in this SPA.

Manager : There is spa massage, sauna, gym, and indoor swimming pool in our establishment. The sauna and gym are open from 7 am to 8 pm. The indoor swimming pool opens at 8 am to 7 pm. The spa massage is open from 10 am to 9 pm. If you want a massage, you should book it 3 hours in advance.

Customer : I got it. Could you tell me the SPA program and the charges?

Manager : The massage courses range from hydrotherapy for 20 minutes at 60,000 won to aroma therapy for 2 hours at 250,000 won.

Customer : What is a hydrotherapy? And, do you have other SPA therapies?

Manager : Hydrotherapy massages the entire body with water. And foot care treatment massages the sole of your foot with a pressure stick. Would you like to try the any massage?

Customer : Thanks for your detail explain.

Manager : Sunny, would you like to make a reservation for a massage?

Customer : Yes, I want to make a reservation for foot care.

Manager : What time do you prefer for a massage? The best time is before a meal and 30 minutes after a meal.

Key words

cancel : 취소하다
correct : 맞는, 정확한, 옳은
detail : 상세, 세부
facilities : 시설
front desk : 프런트, 접수계
meal : 식사
sole : 발바닥, 단하나의, 유일한

4. 스파 예약

매니저 : 안녕하십니까? 샤론 스파입니다. 무엇을 도와 드릴까요?

고객 : 안녕하세요? 써니라고 합니다. 이 호텔 스파에서 어떤 시설을 이용할 수 있는지 알고 싶습니다.

매니저 : 저희 호텔에는 스파 마사지, 사우나, 체력단련실과 실내 수영장이 있습니다. 사우나와 체력단련실은 오전 10시부터 오후 8시, 수영장은 오전 9시 부터 오후 8시까지 사용 가능합니다. 마사지는 오전 9시부터 오후 9시까지 받으실 수 있습니다.

고객 : 아 그래요. 스파 프로그램과 비용을 알려 주시겠습니까?

매니저 : 마사지는 하이드로테라피 20분에 60,000원부터 아로마테라피는 2시간에 250,000원 코스가 있습니다.

고객 : 하이드로테라피가 무엇입니까? 그리고 그 밖에 다른 마사지가 있습니까?

매니저 : 하이드로테라피는 물을 이용하여 전신을 마사지하는 프로그램입니다. 그 밖에 발관리는 지압봉을 이용하여 발바닥에 자극을 주고 풀어주는 마사지입니다. 마사지를 받아 보시겠습니까?

고객 : 설명 감사합니다.

매니저 : 고객님! 마사지를 예약해 드릴까요?

고객 : 네, 발마사지로 예약해 주세요.

매니저 : 마사지를 몇 시에 받으시겠습니까? 참고로 마사지는 식사 전과 식사하고 30분 이후가 좋습니다.

Customer : Okay, I do like reservation at 2 pm.

Manager : I would like to confirm the reservation. Your name is Sunny. The foot care massage will be at 2 pm. Is all this information correct? Foot care massage will take 40 minutes. If your plans change, you can cancel the massage before your scheduled appointment 1 hour before.

Customer : All right!

Manager : Please try to arrive 10 minutes early, so the staff can explain the facility to you. If you want to use the Sauna, please come earlier than your reservation time. And if you need any help, dial 9 or the front desk.

Customer : I see, Bye!

Manager : Thank you! See you at 2 pm, Sunny.

고객 : 그러면 오후 2시가 좋겠습니다.

매니저 : 예약을 확인해 드리겠습니다. 오후 2시 써니님 성함으로 발마사지가 예약되었습니다. 발마사지는 40분이 소요됩니다. 그리고 예약 취소는 1시간 전까지 가능합니다.

고객 : 알겠습니다.

매니저 : 예약 시간 10분 전에 오시면 저희가 간단히 시설을 안내해 드리겠습니다. 감사합니다. 사우나를 이용하시려면 예약시간 전에 시간을 맞춰 이곳으로 오시면 됩니다. 도움이 필요하시면 언제든지 구내번호 9번 혹은 프런트 데스크로 연락 주십시오.

고객 : 알겠습니다.

매니저 : 감사합니다. 오후 2시에 뵙겠습니다.

Reference

- May I have your name and room number?
  제게 성함과 방 번호를 알려 주시겠어요?
- What name is on the reservation?
  어느 분 성함으로 예약하셨습니까?
- The back massage services your waist, back, shoulder, neck and scalp.
  등마사지는 허리, 등, 어깨, 뒷목, 두피 관리입니다.

CHAPTER 03

Makeup Care

1. Telephone inquiry

Manager : Good morning, this is Grace. How may I help you?

Customer : Can I make an appointment over the phone?

Manager : (**When possible**) Sure, you can. When would you like to schedule a visit?

(**When impossible**) I'm sorry. We only set up makeup appointments in person.

Customer : Could you tell me your hours? What days are you closed?

Manager : Sure. We are open from 10 am to 8 pm Monday through Saturday. We are closed on Sundays.

Customer : OK, I' d like to visit you now. How do I get there?

Manager : Where are you coming from?

Customer : I' m at Sejong Center now.

[On Foot]

Manager : Can you see the Kyobo building there?

Customer : Oh, yes

Manager : Next to the building you can see an alley. Go down the alley for about 80 meters and you' ll see a small fancy store "Kitty". There you turn left and you can see the Dream building. We are on the 2nd floor of the building.

Customer : Thank you. I' ll be there soon.

Key words

afraid : 두려워 하는, 걱정 되는
convenient : 편리한
direction : 방향
parking lot (미) 주차장

제3장 메이크업 관리

1. 전화 문의

매니저 : 안녕하세요? 매니저 그레이스입니다.

고객 : 행사에 참석할 메이크업을 하고 싶은데, 전화로도 예약이 가능한가요?

매니저 : 〈**가능할 때**〉 네. 가능합니다. 예약해 드릴까요?

〈**불가능할 때**〉 죄송합니다. 메이크업의 예약은 방문 예약을 원칙으로 하고 있습니다.

고객 : 그렇다면, 영업 시간과 휴무일은 언제인가요?

매니저 : 네, 저희 뷰티 살롱은 일요일은 쉬고 나머지 요일은 영업합니다. 오전 10시부터 오후 8시까지 오시면 됩니다.

고객 : 그 곳으로 찾아가려면 어떻게 가야 하나요?

매니저 : 실례지만 계신 곳이 어디시죠?

고객 : 세종문화회관에 있는데요.

매니저 : 어떤 교통편으로 오실 건가요?

[걸어서 올 때]

매니저 : 교보 빌딩이 보이세요?

고객 : 네

매니저 : 다음 빌딩을 보시면 골목이 있습니다. 골목길을 따라 약 80미터 내려가면 '키티' 라는 작은 팬시점이 보일 겁니다. 그러면 왼쪽으로 돌면 드림빌딩이 있습니다. 그 빌딩 2층에 있습니다.

고객 : 감사합니다. 곧 그곳으로 가겠습니다.

Check point

샴푸 시술과 모발 관리는 헤어 스타일링을 하기 위한 가장 기본적인 준비과정이며, 고객에게 살롱의 차별화된 고품격 서비스를 제공하여 신뢰감과 유대관계를 형성하는 경영전략이 될 수 있다. 또한 전문적인 샴푸 테크닉 시술을 통해 고객의 모발과 두피를 깨끗하게 관리할 수 있게 하며, 예의 바른 접객 멘트를 통해 편안하고 더욱 만족스러운 시술을 제공할 수 있다.

[By Bus or Subway]

Customer : I' m taking a bus/subway.

Manager : Take bus number 88 bound for Chongno, and get off at Chongno 1-ga. The Dream building is right across the street. We are on the 2nd floor. If get lost, call us again.

Manager : (Subway) Get off at Chongno 1-ga, take exit number 8, and walk straight. You' ll see the Dream building on your right. You can't miss it.

[Self-driving]

Manager : Are you self-driving or taking public transportation?

Customer : I'm driving.

Manager : At first come to Chongno 1 ga. You know the Kyobo building?

Customer : Sure.

Manager : Make a turn at the Kyobo building crossroad in the direction of Chunggecheon. On your right side you' ll find our 5-story pink building, the Dream building. We are on the 2nd floor of it.

Customer : Thank you. By the way, can I park my car there?

Manager : (**When possible**) Yes, you can park here any time. Just take your car and park in the building parking lot. We'll be waiting for you.

(**When impossible**) I' m afraid we don' t have a parking lot. But there' s a toll parking lot near our building and you can park there with no charge when you show them our receipt.

Customer : I see. Thank you.

Manager : Thanks for calling. Have a great day.

[버스나 지하철을 탈 때]

고객 : 저는 버스를 탈 겁니다.

매니저 : 종로 방향으로 가는 88번 버스를 타시고, 종로 1가에서 내리세요. 오른쪽으로 건너시면 드림빌딩이 있고, 저희 샵은 2층에 있습니다. 만약 못 찾으시면 다시 한번 전화 주세요.

매니저 : (지하철) 종로 1가역에서 내리시고, 8번 출구로 나와서 똑바로 걸으세요. 그러면 오른쪽으로 드림빌딩이 보일 겁니다. 찾기가 쉽습니다.

[승용차로 올 때]

매니저 : 자가 운전하십니까? 또는 대중교통수단을 사용하십니까?

고객 : 승용차로 가려고요.

매니저 : 제일 먼저 종로 1가를 찾으세요. 교보 빌딩을 아세요?

고객 : 네.

매니저 : 교보빌딩 사거리에서 청계천 방향 회전하세요. 그러면 오른쪽으로 저희 5층의 핑크색 건물, 드림 빌딩을 발견 할 수 있으실 겁니다. 그 건물 2층에 있습니다.

고객 : 감사합니다. 그런데, 거기에 주차 할 수 있을까요?

매니저 : 〈**가능 할 때**〉 네, 언제든 주차하실 수 있습니다. 편안하게 차를 가지고 오셔서 건물 주차장에 주차하시면 됩니다. 그럼 기다리고 있겠습니다.

〈**불가능할 때**〉 죄송합니다만 주차장은 준비되어 있지 않습니다. 근처에 유료 주차장은 있습니다만, 설명해 드릴까요?

고객 : 네, 잘 알았습니다. 감사합니다.

매니저 : 네, 감사합니다. 좋은 하루 되십시오.

Part Ⅲ

Care & Consultation

Business English for Beauticians

CHAPTER 01

Hair

1. Shampooing

Beautician : Hi. I am a hair assistant, Jane. I'll help you out today.

Customer : Oh, nice to meet you, Jane.

Beautician : Pleased to meet you, Mrs. Robinson. Would you come with me to the shampoo room?

Customer : Sure.

Beautician : Please have a seat here. Could you lie on your back slowly? Are you comfortable here?

Customer : My neck is a little bit uncomfortable.

Beautician : Oh. I' m sorry. I' ll lower the chair. Are you all right now?

Customer : Yes. Could you cover my face with a towel?

Beautician : Sure. Let me rinse your hair now. How is the temperature of the water?

Customer : Good, thanks.

Beautician : I' ll apply a shampoo solution. Close your eyes. Is everything okay?

Customer : Actually, my forehead feels itchy. Could you scrub harder, please?

Beautician : Okay. Is there any place that I' ve missed?

Key words

assistant : 보조
comfortable : 편안한
forehead : 이마
help out : 거들다, 원조하다
itchy : 가려운
lie on : ~에 눕다
lean on ..에 기대다
lukewarm : 미지근한
miss : 빠트리다.
patience : 인내, 참을성
scrub : 박박 문지르다
shampoo solution : 샴푸액

제1장 머리관리

1. 샴푸 시술

미용사 : 안녕하세요. 저는 샴푸 도우미 제인입니다. 오늘 고객님을 도와드리게 됐습니다.

고객 : 반가워요, 제인.

미용사 : 반갑습니다. 고객님, 샴푸실로 모시겠습니다.

고객 : 네.

미용사 : 여기 앉아 주십시오. 뒤로 천천히 누워 주시겠어요? 편안하십니까?

고객 : 목이 약간 불편하네요.

미용사 : 죄송합니다. 의자를 조금 낮춰드리겠습니다. 이제 괜찮으신가요?

고객 : 네, 수건으로 얼굴을 좀 가려 주시겠어요?

미용사 : 그러겠습니다. 이제 머리에 물을 묻히겠습니다. 물 온도는 어떠신가요?

고객 : 아주 좋아요. 고마워요.

미용사 : 이제 샴푸를 바르겠습니다. 눈을 감으십시오. 불편한 데는 없으신가요?

고객 : 사실 앞머리 부분이 가렵습니다. 좀 더 세게 문질러 주시겠어요?

미용사 : 네. 혹시 제가 빠트린 부분이 있습니까?

Check point

샴푸 시술과 모발 관리는 헤어 스타일링을 하기 위한 가장 기본적인 준비과정이며, 고객에게 살롱의 차별화된 고품격 서비스를 제공하여 신뢰감과 유대관계를 형성하는 경영전략이 될 수 있다. 또한 전문적인 샴푸 테크닉 시술을 통해 고객의 모발과 두피를 깨끗하게 관리할 수 있게 하며, 예의 바른 접객 멘트를 통해 편안하고 더욱 만족스러운 시술을 제공할 수 있다.

Key words

warm water : 따뜻한 물
hot water : 뜨거운 물
cool water : 시원한 물
cold water : 차가운 물

Customer : No, it' s perfect. Thanks.

Beautician : I' ll rinse your hair now. How is the water temperature?

Customer : It' s too lukewarm.

Beautician : How do you feel now?

Customer : It' s fine. Thanks.

Beautician : It' s done. I' ll dry your hair on this towel. Thank you for your patience.

고객 : 아니요. 아주 좋습니다. 고마워요.

미용사 : 머리를 이제 헹궈드리겠습니다. 물 온도는 어떠십니까?

고객 : 너무 미지근하네요.

미용사 : 지금은 어떠십니까?

고객 : 이제 괜찮아요. 고마워요.

미용사 : 다 끝났습니다. 수건으로 말려드리겠습니다. 머리 감으시느라 수고하셨어요.

2. Hair Treatment

Hair dresser : Good morning, Mrs Parker. How are you today?

Customer : Fine, thank you. I' d like to have a consultation for my hair treatment today.

Hair dresser : Okay. Have a seat here, please. Let me see your hair. Oh, your hair and scalp are a little dry and dull. When was the last time you had any chemical service?

Customer : Several days ago, I got my hair dyed. I feel my hair gets dry and has split ends also. That' s why I want to have my hair treated.

Hair dresser : After applying chemicals to your hair, you should use a special treatment product. Have you used one before?

Customer : Not really. I have been using a special shampoo for damaged hair. Can you recommend a good hair care product for me?

Hair dresser : Of course. I' ll show you some hair treatment products and recommend you a special hair care program. I' m sure your hair will become much more glossy after the treatment.

Customer : I hope so. Thank you for your advice.

Hair dresser : My pleasure.

Key words

advice : 충고, 조언
apply A to B : A 를 B에 바르다
chemical : 화학적인
consultation : 상담
damaged hair : 손상된 머리칼
dyeing, coloring, pigmentation : 염색하다
split ends : 갈라진 끝
recommend : 추천하다
glossy : 반짝거리는, (머릿결이) 매끄러운

2. 모발 관리

미용사 : 안녕하세요, 고객님. 오늘 기분은 어떠세요?

고객 : 좋습니다. 오늘은 모발관리에 대해 상담을 받고 싶습니다.

미용사 : 네, 여기 앉으세요. 모발을 한번 살펴보겠습니다. 아, 고객님 모발과 두피가 좀 건조하고 생기가 없네요. 언제 마지막으로 화학적 시술을 받으셨나요?

고객 : 며칠 전에 모발 염색을 했는데 내 생각에 모발이 매우 건조하고 또한 끝이 많이 갈라졌어요. 그래서 모발 관리를 받아 보려고요.

미용사 : 화학시술을 하신 이후에는 꼭 특별 관리제품을 사용해야 합니다. 전에 사용해 보신 적 있으신가요?

고객 : 아니오. 그냥 손상모발용 특별샴푸를 사용해왔습니다. 제게 맞는 좋은 모발관리 제품을 추천해 주시겠어요?

미용사 : 네, 몇몇 모발 관리 제품들을 보여드리고, 저희 살롱에 있는 특별 모발 관리 프로그램을 추천해 드리겠습니다. 관리를 받으신 후에는 분명히 고객님 머릿결이 훨씬 더 매끄러워질 겁니다.

고객 : 그러면 정말 좋겠네요. 상담해 주셔서 감사합니다.

미용사 : 천만에요.

Key words

Hair analysis parameters

① density
② texture
③ wave pattern
④ hairlines
⑤ grow pattern

3. Dandruff Control Care

Hair dresser : Hello. Thanks for coming in today, despite this weather.

Customer : Yes. It' s very windy too.

Hair dresser : I will keep your umbrella. You have a 3 o' clock appointment. Let me take your bag.

Customer : OK, thanks. But I have a lot of dandruff recently and I am uncomfortable with that. Will it be fine if I get a treatment today?

Hair dresser : For dandruff control, you need one week of special treatment.

Customer : Also, I have an appointment at 6 pm. Will we be finished by then?

Hair dresser : Yes, I will prepare right away. Have you shampooed today?

Customer : Yes. I used a shampoo and a rinse specially designed for damaged hair. I bought them here.

Hair dresser : Scalp treatment kit is on your left. Can you take a seat there?

Customer : Yes.

Hair dresser : First I will check your dandruff condition with the kit and next I' ll take proper care of your hair.

Key words

appointment : 임명, 약속
by then : 그때까지
dandruff : 비듬
damage : 손상
despite : ..에도 불구하고
kit : 장비 일습
prepare : 준비하다
proper : 올바른, 적절한
recently : 최근에

3. 비듬 관리

미용사 : 안녕하세요? 날씨도 안 좋은데 오시느라 고생 많으셨습니다.

고객 : 네. 바람까지 심하게 부네요.

미용사 : 우산을 받아 드릴게요. 3시 예약이시죠? 가방도 맡아드리겠습니다.

고객 : 네, 고맙습니다. 요즘 비듬이 많이 생겨 불편해요. 오늘 관리를 받으면 좋아질까요?

미용사 : 비듬 관리는 일주일 정도 특별 관리를 받으셔야 합니다.

고객 : 여섯 시에 약속이 있는데 그때까지는 끝낼 수 있을까요?

미용사 : 네, 즉시 준비하겠습니다. 오늘 샴푸 하셨습니까?

고객 : 네. 지난 번에 여기서 구입한 손상 모발용 샴푸와 린스를 사용했어요.

미용사 : 왼쪽에 두피관리기기가 있습니다. 그쪽으로 앉으십시오.

고객 : 네.

미용사 : 우선 두피관리기기를 이용하여 비듬상태를 확인한 다음에 적절한 관리를 실시하겠습니다.

Key words

Hair analysis parameters

① density
② texture
③ wave pattern
④ hairlines
⑤ grow pattern

4. Hair Cut & Styling

Hair dresser : Hi. I am a hair designer, Kelly. You made an appointment for a hair cut today, Ms. Pore.

Customer : Yes, nice to meet you.

Hair dresser : Nice to meet you, too. Can you tell me what kind of style you want ?

[Case I]

Customer : I need a long style easy to manage because i do not have much time in the morning.

Hair dresser : Would you like to look at some style books?

Customer : Yes, please. How about this style?

Hair dresser : It's a easy to manage style, but I am afraid it wouldn' t look good on you. I think it' s better to make lots of volume around your chin, because you have an oblong face.

Customer : How about this one? Will this style be suitable for me?

Hair dresser : Yes, it' s a good choice. This style should give you a soft look. It will make your hair feel light and be easy to manage for yourself.

Customer : Sounds great. Let' s do that.

Hair dresser : You want to keep your hair long, don' t you?

Customer : A little past my shoulder.

Hair dresser : I also think shoulder length will look good on you.

Key words

bet : 내기, 생각
예) **Your best bet is to apologize.** 사과하는 것이 상책이다
blow dry : 말려서 드라이하다
chin : 턱
congratulations : 축하합니다
cut off 잘라내다
look good on : 잘 어울리다
loose : 헐거운, 느슨한, 풀린
major : 대부분의, 주요한
oblong : 장방형의
past : ~를 지나서
particular : 특별한, 특정의
round : 원형의, 둥근
stand out : 돋보이다, 특출나다
suitable : ~에 적합한
trim : 손질, 정돈
volume : 부피, 분량
wavy : 곱슬거리는, 컬이 있는
for yourself : 혼자서
trim : 손질하다, 정리하다, 잘라내다

4. 커트 & 스타일링

미용사 : 안녕하세요. 저는 헤어 디자이너 켈리입니다. 오늘 머리 커트 예약 하셨죠?

고객 : 네, 맞습니다. 만나서 반가워요.

미용사 : 만나게 되어 저도 반갑습니다. 오늘 어떤 스타일을 하고 싶으세요?

[경우 1]

고객 : 저는 아침마다 시간이 많지 않아서 그러는데 관리하기 쉬운 긴 머리 스타일을 하고 싶은데요.

미용사 : 여기 스타일 사진집 좀 보시겠어요?

고객 : 네 그러죠. 이 스타일은 어떤가요?

미용사 : 이 스타일은 집에서 관리하기 쉽긴 한데 고객님께 어울리진 않을 것 같습니다. 고객님은 얼굴이 장방형이라 턱 주위에 볼륨감을 줄 수 있는 스타일이 좋겠습니다.

고객 : 그럼 이 스타일은 어떤가요? 저에게 적당한가요?

미용사 : 네, 탁월한 선택입니다. 이런 스타일은 얼굴형을 부드럽게 보이게 하지요. 또한 모발이 가벼워 보이게 하고, 혼자 관리하기도 편하실 거예요.

고객 : 괜찮은데요. 이 스타일로 하겠어요.

미용사 : 머리를 계속 기르실 건가요?

고객 : 어깨 길이 정도로요.

미용사 : 제 생각에도 고객님께는 어깨 길이가 잘 어울릴 거 같아요.

Customer : Kelly! I want to blow dry for a special occasion after my haircut. Actually, I have a party tonight. Could you do that for me?

Hair dresser : Sure, what kind of party are you having?

Customer : It' s my birthday party.

Hair dresser : Oh, really? Congratulations! Do you have any particular style that you want?

Customer : A loose wavy and elegant style.

Hair dresser : No problem. I' m going to make you stand out today.

Customer : Thank you.

Hair dresser : You' re welcome.

[Case II]

Customer : I still have a lot of curl left in my hair, but I have ajor split ends. I want my hair trimmed. Even so, I want to keep my hair long too. How many inches do you suggest?

Hair dresser : I would say your best bet is 2 inches. And if that is not enough then we can go a little shorter. We always start with less because you can always cut more off but you can' t put it back on. Just cut enough to make it healthy the only way to get rid of split ends is to cut them off.

Customer : I know after a good trim, a lot of the old perm will be cut off. I want a new perm too.

고객 : 머리를 자른 후에 특별한 자리에 어울리는 스타일로 드라이하고 싶은데요. 사실 오늘 저녁에 파티가 있거든요. 해주실 수 있겠어요?

미용사 : 물론이죠. 무슨 파티에 가시나요?

고객 : 제 생일 파티예요.

미용사 : 어머, 정말이세요? 진심으로 생일 축하드려요. 특별히 하고 싶은 스타일 있으세요?

고객 : 약간 곱슬거리면서 우아한 스타일로 하고 싶어요.

미용사 : 문제 없습니다. 오늘 파티에서 돋보이실 수 있도록 해드릴게요.

고객 : 고마워요.

미용사 : 천만에요.

[경우 2]

고객 : 아직도 머리가 많이 곱슬거리기는 한데 머리끝이 많이 갈라졌어요. 그래서 머리 를 가지런히 다듬어 주셨으면 합니다. 그래도, 긴 머리는 계속 유지하고 싶어요. 몇 인치 정도 자르면 좋을까요?

미용사 : 가장 좋은 것은 2인치라고 할 수 있죠. 만약에 충분치 않다고 생각되시면 좀 더 짧게 자를 수도 있고요. 언제나 처음에는 좀 덜 자르는데, 왜냐하면 좀 더 자를 수는 있지만 다시 덧붙일 수는 없잖아요. 머리카락을 건강하게 유지하기 위해서는 갈라진 끝을 넉넉히 자르세요. 끝이 갈라진 머리카락을 제거하는 유일한 방법은 잘라내는 방법밖에 없습니다.

고객 : 머리를 자르고 나면 펌 머리가 많이 잘려나갈 거예요. 그래서 펌도 다시 해야겠어요.

Useful abbreviation of hair point

Abbreviation of hair point (두부의 부위별 약자)

Abbreviation	Point
E.P	Ear Point
C.P	Center Point
T.P	Top Point
G.P	Golden Point
B.P	Back Point
N.P	Nape Point
F.S.P	Front Side Point
S.P	Side Point
S.C.P	Side Corner Point
E.B.P	Ear Back Point
N.S.P	Nape Side Point
C.T.M.P	Center Top Medium Point
T.G.M.P	Top Golden Medium Point
G.B.M.P	Golden Back Medium Point
B.N.M.P	Back Nape Medium Point

Parting of scalp(두부의 분할)

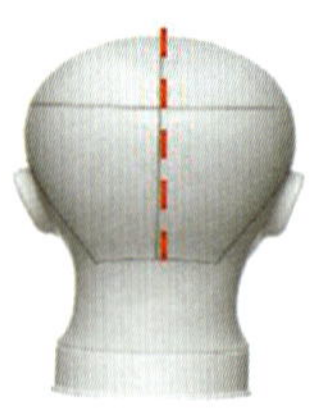	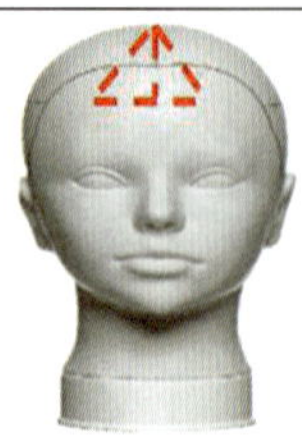	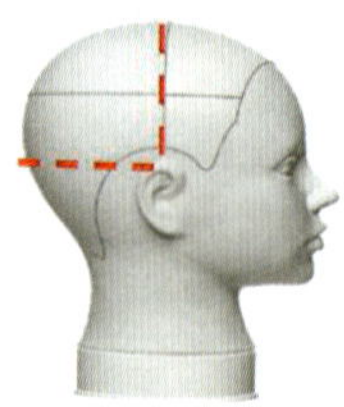	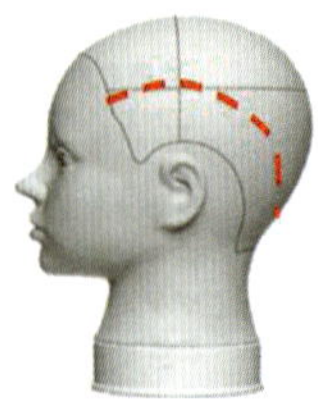
Center part	Side part	Ear to Ear part	Cure part

Kinds of basic hair cut (기본 커트의 종류)

1) solid cut (one length cut) : ① straight Bob cut ② Isadora cut ③ Spaniel cut
2) graduation cut : ① low graduation cut, ② medium graduation cut, ③ high graduation cut
3) increase layer cut : ① vertical increasing layer cut, ② forward increasing layer cut, ③ horizontal increasing layer cut
4) uniform layer cut

Hair layer (모발 층)

① jagged
② progressive
③ razored
④ stacked

Hair style/shape (모발 스타일/형태)

backcombing	finger tousling
bang	fringe
blunt cut	jagged
bob	lock/tress
braided	pig tail
bun	pony tail
crew cut	reggae
crimps	shaggy
crop	tousled

5. Perm & Coloring

(1) Perm wave Ⅰ

Hair dresser : Hi. It' s been a long time, Miss Judy.

Customer : How have you been?

Hair dresser : Very well. How are you? You had an appointment for a perm today.

Customer : Right.

Hair dresser : What type of perm would you like to try today?

Customer : You know that I' ve always had straight hair. I' d like to change my style in vogue and also look natural.

Hair dresser : Oh, I see. Would you like to see pictures of the popular style?

Customer : Yes, please. The first pic on this page might work for me, loose and spiral curls. What do you think about that?

Hair dresser : Miss Judy. I think that' s a perfect choice. This style looks natural and that is also modern and popular these days. I' m sure it will be good on you.

Customer : Oh, I' m so excited to see what it'll look like.

Hair dresser : All right, let me prepare the things for your perm. Please wait a second and relax.

Customer : Okay, take your time.

Hair dresser : I' ll first wind the perm rods and then I' ll apply the perm solution. If you are uncomfortable, at anytime just let me know.

Key words

popular : 유행하는 (= in style, in vogue)
pic = picture
spiral : 나선형의
perfect choice : 탁월한 선택
perm rod : (퍼머용) 작은 막대
wind〔waind〕 : 말다, 감다 ↔unwind
winding : (rod)감기
being pulledz : 잡아당기다
neutralizer : 중화제
fix : 고치다
stay in : 유지, 남아있다
drip on : ~에 떨어지다
clean up : 깨끗이 하다
towel dry : 수건으로 머리를 닦다/말리다
flip out : ~가 밖으로 뻗치다
maintain : 관리하다 (= manage)

5. 펌과 염색

Check point

화학적인 시술인 펌 웨이브와 염색은 고객의 안전에 대한 세심한 주의와 모발 건강을 위한 시술 전후 모발 케어가 무엇보다 중요하다. 펌 스타일과 염색에 대한 사전 상담을 통해 모발의 상태를 점검하고, 화학 시술 시 주의깊은 고객 관리와 시술 과정에 대한 설명 및 정보 전달을 통해 신뢰와 편안함을 제공해야 한다.

(1) 펌 웨이브 시술 I

미용사 : 안녕하세요? 오랜만입니다, 주디양.
고객 : 어떻게 지냈어요?
미용사 : 덕분에 잘 지냈습니다. 오늘 펌 예약 하셨죠?

고객 : 네.
미용사 : 오늘 어떤 스타일의 펌을 하고 싶으십니까?
고객 : 아시겠지만, 제가 그동안 늘 생머리만 해왔잖아요. 오늘은 뭔가 유행을 따르면서도 자연스러운 스타일로 바꾸고 싶어요.
미용사 : 그러세요. 유행하는 스타일의 사진들을 한 번 보시겠습니까?

고객 : 네, 좋아요. 이 페이지의 첫 번째 사진이 저랑 잘 어울릴 거 같은데, 약간 늘어지고 꼬불거리는 컬의 이 스타일요. 어때요?
미용사 : 고객님. 정말 잘 고르셨어요. 이 스타일은 자연스러운 동시에 최근 가장 유행하는 스타일입니다. 고객님께 아주 잘 어울릴 거예요.

고객 : 어떻게 보일지 정말 기대되는데요.
미용사 : 좋아요. 펌 웨이브 준비를 하겠습니다. 편안히 기다려 주세요.

고객 : 네, 천천히 준비하세요.
미용사 : 먼저 펌 로드로 말고 그 후에 펌 용액을 도포하겠습니다. 시술 중에 불편하시면 꼭 말씀해 주십시오.

Customer : Okay, thanks.

Hair dresser : I just finished with the winding. I' ll apply the perm solution. Are you OK?

Customer : It feels like the hair behind my ear back is being pulled.

Hair dresser : Oh, I' m so sorry. Let me fix it. Is it still being pulled?

Customer : No. I' m fine now. Thanks.

Hair dresser : I' ll apply neutralizer, it needs to stay in for thirty minutes.

Customer : Okay. Would you give me a tissue? I feel the neutralizer is dripping on my neck.

Hair dresser : I' ll clean it up for you. Are you okay now?

Customer : Yes, I am. Thank you.

Hair dresser : Miss Judy, it's done. I' m unwinding the perm rods. I' ll wash your hair. The wave looks great.

Customer : Oh, I love the way I look. Thanks a lot. Could you tell me how to maintain my hair?

Hair dresser : Towel dry and then apply hair gel from the ends. You can also flip the hair out with hair wax.

Customer : I' m so happy. Thank you for your service today.

Hair dresser : It' s my pleasure.

고객 : 그러죠, 고마워요.

미용사 : 펌 말기가 끝났습니다. 펌 용액을 도포하겠습니다. 괜찮으십니까?

고객 : 제 생각엔 귀 뒤쪽 머리카락이 당기는 것 같아요.

미용사 : 죄송합니다. 제가 다시 말아드리겠습니다. 아직도 당기십니까?

고객 : 아니요. 이제 괜찮아요. 고마워요.

미용사 : 중화제를 도포하겠습니다. 30분 정도 그대로 계세요.

고객 : 알겠어요. 티슈 좀 줄래요? 중화제가 제 목 위로 흐르는 것 같아요.

미용사 : 제가 깨끗하게 닦아 드리겠습니다. 이제 괜찮으십니까?

고객 : 네, 고마워요.

미용사 : 고객님, 다 됐습니다. 로드를 풀고 헹구어 드리겠습니다. 컬이 잘 나왔네요.

고객 : 정말 맘에 들어요. 고마워요. 머리 관리는 어떻게 하는지 알려줄래요?

미용사 : 수건으로 닦아 말리고 젤을 모발 끝에서부터 바르세요. 또 왁스로 모발을 바깥으로 뻗치게 할 수도 있습니다.

고객 : 알겠어요. 정말 마음에 드네요. 오늘 고마워요.

미용사 : 별 말씀을요.

(2) Perm wave II

Hair dresser : Tell me which perm style you want?

Customer : Can you give me a loose and natural wave?

Hair dresser : Yes. But your hair seems to be quite porous and the perm style could not last long. Is it still okay?

Customer : That’ s fine. Could you make it naturally curled?

Hair dresser : Of course. When was the last time you had a perm?

Customer : Two months ago.

Hair dresser : I will give you a perm after the treatment.

Customer : How long does it take?

Hair dresser : About two hours. Is it fine?

Customer : Yes. Please go ahead.

Hair dresser : Did you rinse your hair before you came?

Customer : Yes, I shampooed it.

Hair dresser : OK. I will start your treatment now.

Customer : I have a sensitive scalp. I don’ t want to wear the electrical hat.

Hair dresser : OK. I will use a steamer. Is that alright?

Customer : Yes.

Hair dresser : You’ re welcome. Feel free to tell me if you need anything else.

Customer : OK. Thanks.

Key words

electrical : 전기의
go ahead : 계속하다
last : 지난, 지속되다, 견디다
porous : 작은 구멍이 많은, 투과성의
sensitive : 민감한, 예민한
steamer : 스티머
wear a hat : 모자를 쓰다

(2) 펌 웨이브 시술 II

미용사 : 어떤 스타일의 펌을 원하세요?

고객 : 굵고 자연스러운 웨이브를 하고 싶은데 가능할까요?

미용사 : 가능합니다. 그런데 고객 모발이 다공성이라 그런 스타일은 오래가지 않을것 같은데요. 그래도 괜찮으십니까?

고객 : 괜찮아요. 자연스럽게 곱슬거리도록 해주실 수 있죠?

미용사 : 그럼요. 펌을 마지막으로 한 것이 언제입니까?

고객 : 두 달 전이에요.

미용사 : 머릿결 관리를 한 다음 펌 시술을 해야겠네요.

고객 : 시간은 어느 정도 걸릴까요?

미용사 : 두 시간정도 소요됩니다. 괜찮으세요?

고객 : 좋아요. 그렇게 해주세요.

미용사 : 오시기 전에 린스를 하고 오셨습니까?

고객 : 네, 샴푸했어요.

미용사 : 그러면 지금부터 클리닉을 하겠습니다.

고객 : 두피가 예민하니까 전기 모자는 사용하지 말아주세요.

미용사 : 네. 스티머를 이용할게요. 괜찮으시죠?

고객 : 네. 고마워요.

미용사 : 천만에요. 필요한 것이 있으면 언제든지 말씀해 주십시오.

고객 : 네. 고맙습니다.

(3) Coloring I

Hair dresser : Hi. Have you been doing well?

Customer : Yes, I have. I want to have my hair dyed today. What color do you think will look good on me?

Hair dresser : You' ll look good with a warm color, because you have a slim figure. Would you like to look on the color swatch?

Customer : Yes, please. I' d like to have a brown hue like this.

Hair dresser : That' s a good choice. You' ll look good with this chestnut brown or dark chocolate with blond highlights. Or if you want a light brown color, golden brown, camel brown or light tan will work as well. If you get brownish brunette with fiery orange highlights, it' s a good combination.

Customer : I want to try the chestnut brown with blond highlights.

Hair dresser : Excellent choice. When was the last time you had any chemical service?

Customer : I got my hair permed two months ago.

Hair dresser : All right, it will be fine to dye your hair today. Could you wait a moment? I' ll prepare for coloring.

Customer : Okay. Thank you.

Key words

brunette : 검은색
brown hue : 갈색조
camel brown : 담황갈색
chemical service : 화학적인 시술(파마, 염색)
chestnut brown : 밤색
dark chocolate : 진한 초콜릿 색
dye : 염료, 염색하다
fiery : 불같은
figure : 모양, 체격, 형태
a good combination : 잘 어울리는 배합
golden brown : 금갈색
light : 가벼운, 옅은
light tan : 옅은 황토색
slim : 날씬한
swatch : 견본

(3) 염색 시술 I

미용사 : 안녕하세요. 잘 지내셨어요?

고객 : 네, 잘 지냈어요. 오늘 염색하고 싶은데요. 무슨 색이 저랑 어울리겠어요?

미용사 : 고객님이 좀 마른 체격이라서 따뜻한 색의 모발이 어울릴 것 같습니다. 모발 견본 차트 한 번 보시겠어요?

고객 : 네, 좋습니다. 이 색처럼 갈색 계열로 하고 싶어요.

미용사 : 잘 고르셨어요. 고객님께는 여기 금갈색으로 악센트를 준 밤색 혹은 짙은 초콜릿 색이 잘 어울릴 거예요. 혹은 옅은 갈색을 원하신다면, 금갈색, 담황색 혹은 옅은 황토색도 괜찮고요. 만약 타오르는 주황색으로 악센트를 준 흑갈색을 원하신다면, 그것도 잘 어울리는 배합이 될 겁니다.

고객 : 금갈색 하이라이트가 있는 밤색으로 해보고 싶어요.

미용사 : 잘 고르셨어요. 화학 시술을 받으셨던 마지막 때가 언제입니까?

고객 : 2달 전에 펌을 했습니다.

미용사 : 네, 오늘 머리 염색을 해도 괜찮을 거예요. 조금 기다려 주시겠어요? 염색준비를 하겠습니다.

고객 : 네, 감사합니다.

Reference

Hair sheen

Bouncy	porous/brittle/fragile
chunky	rough
glam=glamour/glamorous	sheen
gloss/glossy	shiny
lustrous, luster	sleek
natural swing	smooth
oily	wiry

(4) Coloring II

Hair dresser : Hello. You look great today.

Customer : Hi. Can you dye my hair?

Hair dresser : Your hair was dyed brown before. What color do you want today?

Customer : Nowadays, I am concerned about looking old. I want myself to look young and neat. What color would be good in this case?

Hair dresser : A bright brown color will complement your skin tone and it will make you look younger.

Customer : How about something other than brown? I want blonde or brunette.

Hair dresser : Emm, how about brownish brunette with straw blonde highlights, or all blonde?

Customer : That sounds good. I'll get brownish brunette with straw blonde highlights.

Hair dresser : That color makes your complexion brighter.

Key words

be concerned about : ~ 대해서 걱정하다
brownish brunette : 갈색빛 나는 검은색
complement : 보충하다
complexion : 안색
nowaday : 요즘의, 오늘날의
other than : ~ 말고, 과는 달리
straw blonde : 밝은 담황색

(4) 염색 시술 II

미용사 : 어서 오세요. 얼굴이 좋아 보이시네요.

고객 : 안녕하세요. 염색해 주실 수 있죠?

미용사 : 전에 브라운 컬러로 염색하셨는데 오늘은 어떤 컬러를 원하십니까?

고객 : 요즘 나이 들어 보여서 고민입니다. 젊고 산뜻해 보였으면 좋겠는데. 어떤 컬러가 좋을까요?

미용사 : 밝은 브라운 색깔이 고객님 피부 톤을 살려줄 거예요. 그리고 더 젊어보이게 할 거고요.

고객 : 그럼 갈색 말고 다른 색은 어때요? 블론드나 검은 머리도 좋아요.

미용사 : 음... 그럼 밝은 담황색 악센트가 있는 흑갈색이나 전체를 블론드로 하면 어떠세요?

고객 : 괜찮아요. 밝은 담황색 악센트가 있는 흑갈색으로 할 거예요.

미용사 : 그 색깔이 고객님의 안색을 더 빛나게 해줄 거예요.

Reference

Hair color

① brown : butterscotch, chestunt brown, cinnamon, coffe-colored, dark/deep/light chocolate, hazelnut brown, milk chocolate, mocha

② yellow : beige, blonde, gold blonde, ivory, mud, staw blondem, tan

③ red : aubum, cardinal, orange, ruby, scarlet

④ pink : cerise

⑤ green : blush green, emerald green, jade green, peridot, turquoise blue, verdure/verdant, yellowish green

⑥ white : ivory, snow white

⑦ blue : aquamarine, azure, deep blue, lapis lazuli, pale blue, sapphire, sky blue

⑧ purple : amethyst, crimson

⑨ etc. : black, brunette, butter scotch, charcoal grey, cinnamon brown, cinnamon spice, cofee-colored strands, dark choclate, deep chocolate, grey, irish creme, milk chocolate, mocka fudge, shades, of brunette, silbery

6. Up-Style

Hair dresser : Are you going to a special party today?

Customer : Yes, I am attending a year-end party tonight.

Hair dresser : Really? Do you have any particular style that you want?

Customer : I' d like to get an up-style to go well with my light purple cocktail dress.

Hair dresser : Let me show you some up-style that may work for you?

Customer : How about this hair style? I think it would look good on me.

Hair dresser : Sure, I agree with you. Let me see. For making up-style voluminous, I'll use some hair pieces.

Customer : Oh, I like it. Please do whatever you think is best.

Hair dresser : I' ll change you into a classic, elegant and sophisticated lady today. It' s going to match your dress well. I' m sure you' ll look fabulous tonight.

Customer : I am so flattered.

Key words

agree with : ~에 동의하다
attend : 참여하다
classic : 고전적인, 고상한
elegant : 우아한
fabulous : 훌륭한, 멋진
flattered : 추켜세워진, 칭찬의 말 때문에 부끄러운
go well with : 잘 어울리다 (= work for, look good on, match well)
hair piece : 부분 가발
light purple : 옅은 자주색
sophiscated : 세련된, 멋진
voluminous : 넉넉한, 부피가 큰, 볼륨감 있는
year-end party : 송년회

6. 업스타일 상담

미용사 : 오늘 특별한 파티에 가시나요?

고객 : 네, 오늘 저녁에 송년회에 갑니다.

미용사 : 아? 그러세요. 특별히 하고 싶은 스타일이 있으십니까?

고객 : 연보라빛 칵테일 드레스랑 잘 어울리는 업스타일로 올리고 싶어요.

미용사 : 고객님에게 어울릴 만한 헤어스타일 사진을 찾아보시겠어요?

고객 : 이 스타일은 어떤가요? 제 생각에는 저한테 잘 맞을 것 같아요.

미용사 : 저도 동감입니다. 어디 봅시다. 업스타일을 볼륨감 있게 살려주기 위해서는 저는 부분 가발을 이용할거예요.

고객 : 그거 참 좋겠네요. 멋지게 스타일링 해주세요.

미용사 : 오늘 고전적이고 우아하고 세련된 숙녀로 변신시켜드릴게요. 드레스와 잘 어울려서 오늘 저녁에 분명히 아름다우실 겁니다.

고객 : 감사합니다.

Check point

모임이나 기념일 등의 특별한 날에 헤어 스타일링을 하는 고객이 늘어나고, 파티 문화가 보편화되면서 헤어 업스타일은 웨딩에서 뿐 만 아니라 평소에도 유행하는 헤어 살롱의 메뉴가 되고 있다. 고객의 얼굴형에 맞고 기념일에 잘 어울리는 업스타일을 시술하기 위해서는 고객과 원활한 커뮤니케이션을 나누어야 한다.

CHAPTER 02 Skin Care

1. Skin Care Consultation

Esthetician : Let me brief you on your 90 minute facial treatment. Processes are as follows; cleaning, skin analysis, decollete massage, first mask, extraction, relaxing massage, second massage, finishing touch.

Customer : That sounds wonderful.

Esthetician : After your facial treatment, what would you like to change about your skin?

Customer : I don' t know exactly, but my skin feels heavy.

Esthetician : I see, I think you need to slough off dead skin cells. Are you currently using any treatments for acne? Or have you used deep exfoliation products?

Customer : No. I just use general care products at home.

Esthetician : OK, that' s fine. We can use an exfoliating product. After exfoliating your dead skin, you' ll feel light and clean.

Customer : That sounds great. Thank you.

Key words

- **acne** : 여드름
- **corneous tissue** : 각질조직
- **extraction** : 추출, 적출
- **exfoliate** : 벗겨내다, 박리시키다.
- **dead skin cell** : 각질 (= horny substance, keratin)
- **pill** : 알약, 정제
- **a stratum corneum** : 각질층

제2장 피부관리

1. 피부관리 상담

피부미용사 : 90분 피부 관리 서비스에 대해 간단하게 소개하겠습니다. 관리 진행은 세안, 피부 분석, 해독작용 마사지, 첫 번째 마스크, 추출(과잉피지 및 여드름 제거), 피부 휴식을 도와주는 마사지, 두 번째 마스크, 마무리(기초제품 및 메이크업) 순으로 구성되어 있습니다.

고객 : 좋은데요.

피부미용사 : 피부관리 후 고객님의 피부가 어떻게 개선되길 바라십니까?

고객 : 정확히 잘 모르겠어요. 하지만 피부가 무거운 느낌이 들어요.

피부미용사 : 알겠습니다. 제 생각에는 고객님 피부 상태는 각질 제거가 필요할 것 같은데요. 최근 여드름 약을 복용하시나요? 혹은 각질 제거 제품을 사용하고 계십니까?

고객 : 아니요. 저는 그냥 일반적인 홈케어 제품을 사용하고 있는데요.

피부미용사 : 네. 그렇다면 좋습니다. 각질 제거 제품을 사용할 수 있겠네요. 피부 각질을 제거하면 피부가 가벼워지고 깨끗해진 느낌이 드실 겁니다.

고객 : 좋을 것 같네요. 고마워요.

Check point

피부관리실로 고객을 안내한 후 고객이 받을 시술에 대한 간단한 설명과 고객의 피부 상태를 파악한다. 그 후 개선점에 대한 상담을 실시한다. 고객상담 시 피부미용사는 전문적인 태도로 고객의 신뢰를 얻도록 해야 한다.

2. Preparation

Esthetician : Please, take off any jewelry you may be wearing : rings, bracelets, earrings, watches and necklaces. And place them in this box.

Customer : Ok, thank you!

Esthetician : Please change into this gown and let me know when you' re ready.

Customer : Anna. I am ready. You can come in.

Esthetician : OK, Christine. Please, lie down on the bed with your head on this side.

Customer : OK! Grace.

Esthetician : I am going to cover your body with this sheet. Do you want it tight or loose?

Customer : Tight, please.

Esthetician : Let me put the towel around your hair first so nothing gets into your hair.

Key words

establishment : 영업장소, 사업장
rewarding : 가치 있는
secure : 잘 간수하다
~ the valuables 귀중품을 간수하다
assume : (책임) 맡다
~ responsibility for – 에 대해 책임을 지다
as such : 그러니까
misplace : –를 잘못 두다
assist : 돕다
take off : 벗다
jewelry : 보석류
bracelet : 팔찌
earring : 귀걸이
necklace : 목걸이
place : 놓다
lie down : 눕다
put the towel around – 주위에 타월을 두르다

2. 준비

피부미용사 : 이제, 지금 착용하고 있는 보석류(반지, 팔찌, 귀걸이, 목걸이, 시계)를 빼주십시오. 그리고 보석과 귀중품은 여기 보관함 속에 넣어주십시오.

고객 : 알았습니다. 감사합니다.

피부미용사 : 이 가운으로 갈아입으시고, 준비가 되면 알려 주십시오.

고객 : 안나씨, 준비가 다 되었는데요. 들어오셔도 돼요.

피부미용사 : 고객님, 머리를 이쪽으로 하고 누워 주십시오.

고객 : 알겠습니다.

피부미용사 : 이제 시트로 몸을 덮어드리겠습니다. 단단히 감쌀까요? 아니면 느슨하게 할까요?

고객 : 단단하게 싸주세요.

피부미용사 : 머리카락 사이로 아무것도 들어가지 못하게 먼저 머리를 타월로 두르겠습니다.

3. Makeup Removal

Esthetician : I' m going to clean your face with a cream cleanser. Do you wear contact lenses?

Customer : No, I don' t.

Esthetician : OK, now I will use eye and lip makeup remover. Please, close your eyes, I' m going to stroke down your eyelids and eyelashes. And then look up for me, so I can remove the makeup underneath your eyes.

Customer : OK.

Esthetician : Let me apply makeup remover on your lips. Please, close your lips. Okay, I will wash off your eye and lip makeup. After I apply the cream cleanser, I will rub it over your face. Next I' ll remove the cleanser. I' ve finished the cleansing process. Now we will begin the next step.

Customer : Thank you.

Esthetician : Next I' m going to apply the toner. This product will be sprayed directly onto your face and it might be a little cold. Please close your eyes tightly.

Key words

proceed : 계속하다, 처리하다
eyelash : 눈썹
makeup remover : 화장 지우기용 크림
product : 물품
rub : 문지르다, 맞비비다
stroke down : 쓸어내리다
underneath : 아래의, 아래에

3. 화장지우기

피부미용사 : 크림 타입 클렌저를 이용하여 얼굴 화장을 닦아드리겠습니다. 혹시 콘택트렌즈를 착용하셨나요?

고객 : 아니요.

피부미용사 : 그럼 눈과 입술 화장 리무버를 사용하여 닦아내겠습니다. 눈을 감아주세요. 먼저 눈꺼풀과 눈썹 부분을 아래로 닦아내겠습니다. 그런 다음 눈을 위로 떠 저를 보세요. 눈 아래 부분의 메이크업을 지우겠습니다.

고객 : 알았습니다.

피부미용사 : 입술에 리무버를 바르겠습니다. 입을 다물어 주세요. 눈과 입술 부분의 메이크업을 지우겠습니다. 크림 타입의 클렌저를 얼굴 전체에 바르겠습니다. 다음으로 클렌저를 닦아내겠습니다. 화장지우기 과정이 끝났습니다. 다음과정을 계속 진행하겠습니다.

고객 : 고맙습니다.

피부미용사 : 토너를 사용하겠습니다. 이 제품은 얼굴에 직접 분사하는 스프레이 형으로 차가울 수 있습니다. 눈을 꼭 감아주세요.

4. Facial Skin Analysis

Esthetician : What are you mainly concerned about with your skin?

Customer : Um, I often have dry skin.

Esthetician : I see. Let me examine your skin before proceeding with your treatment. Please, close your eyes. I' m going to cover your eyes with eye pads. I will use a Wood' s lamp. It might be a little bright.

Customer : No problem. Could you tell me about my skin condition?

Esthetician : Of course, I will analyze your skin type, tone and texture.

Key words

broken : 깨진, 부러진
capillary : 모세혈관
comedo : 면포 단계의 여드름
concern : 걱정하다
distend : 팽창하다
encrustation : 피각, 각질, 외피
even : (형) 평평한, 고른, (명) 걱정, 근심
gorgeous : 눈부신, 훌륭한
omit : 생략하다, 빼다
procedure : 순서, 진행
a sebaceous gland : 피지선
stimulate : 자극하다

4. 안면 피부 분석

피부미용사 : 고객님의 피부 고민은 무엇입니까?

고객 : 음, 피부가 조금 건조해요.

피부미용사 : 알겠습니다. 피부관리를 시행하기 전에 피부를 살펴보겠습니다. 자, 눈을 감아주세요. 아이패드로 눈을 덮겠습니다. 그리고 우드램프를 사용할 텐데 눈이 부실 수 있습니다.

고객 : 괜찮아요. 제 피부 상태에 대해 설명해 주실거죠?

피부미용사 : 물론이죠. 고객님의 피부 유형과 피부톤, 피부결을 분석할 겁니다.

Dry skin type : Case 1

Esthetician : First, your skin is a dry skin type. Second, your face' s skin temperature is the same as your body. Your tone is even but the texture is a little rough. After washing your face, how does the skin feel on your face?

Dry skin type : Case 2

Esthetician : In your case, oil balance is important. Facial treatment should help stimulate the sebaceous glands and normalize the production of sebum. So I will apply a mask and massage for dry skin.

Normal skin type : Case 1

Esthetician : Your skin is absolutely gorgeous, you have normal skin. Your skin tone is even and smooth, but your temperature of your face is higher than your body.

Normal skin type : Case 2

Esthetician : Your skin is in good condition, having enough sebum and moisture. If you treat your skin regularly, you can keep good skin continuously.

건성 피부 유형 : 예 1

피부미용사 : 먼저 고객님의 피부 유형은 건성입니다. 두 번째, 얼굴의 피부온도가 몸의 온도와 같고 피부 톤도 고르지만, 피부결이 약간 거칠어진 것 같습니다. 세안을 하신 후에는 피부가 어떤가요?

건성 피부 유형 : 예 2

피부미용사 : 고객님의 경우에는 피지 분비의 밸런스가 중요합니다. 그래서 피지샘을 자극하고 피지 생성을 정상화할 수 있는 피부 관리가 되도록 하겠습니다. 건성피부를 위한 마스크와 마사지를 적용하겠습니다.

중성 피부 유형 : 예 1

피부미용사 : 피부 상태가 매우 좋으며 중성 피부입니다. 피부톤도 고르고 매끄러우나 피부 온도가 체온보다 약간 높습니다.

중성 피부 유형 : 예 2

피부미용사 : 피지와 수분이 충분해 피부 상태가 매우 좋은 정상 피부입니다. 꾸준히 관리를 잘 해 주시면 좋은 피부를 지속적으로 유지하실 수 있습니다.

Normal skin type : Case 3

Esthetician : You already have great skin. So you should keep your skin condition moistened with facial treatment and home maintenance. I will give you a mask and massage. One more thing, your skin temperature is a little bit high. I think this is a temporary condition. So today' s treatment will focus on bringing your temperature down.

Combination skin type : Case 1

Esthetician : Your skin is a combination skin type. Your skin is oily and dry. I mean your T-zone, especially your nose and brow area is oily, however, your forehead and cheek area is dry.

Combination skin type : Case 2

Esthetician : In my opinion, your skin type is complex. Your skin appears both greasy and dry : the T-zone of your nose and brow is glossy because of much sebum secretion, but you feel dryness on your forehead and cheeks.

Combination skin type : Case 3

Esthetician : I would like to treat your balanced skin condition. I am going to apply two different masks, one for dry skin and the other for oily skin. I will apply a dis-encrustation mask on your T-zone, and an intensive hydrating mask on your forehead and cheeks.

중성 피부 유형 : 예 3

피부미용사 : 피부가 좋으네요. 피부 마사지나 집에서 꾸준한 관리를 통해 피부 상태를 유지하는데 중점을 두셔야 합니다. 제가 마스크와 마사지를 시행하겠습니다. 한 가지 더 말씀드리고 싶은 것은 고객님의 얼굴 피부 온도가 약간 높은데, 제 생각에는 일시적인 현상인 것 같습니다. 그래서 오늘 관리는 피부 온도를 진정시키는 것에 초점을 맞추겠습니다.

복합성 피부 유형 : 예 1

피부미용사 : 제 생각에는 고객님의 피부는 복합성 피부에 해당합니다. 고객님의 피부는 지성과 건성이 모두 나타나는데 즉 고객님의 얼굴 중 특히 코와 미간의 T존 부위는 지성이고, 이마와 볼은 건성입니다.

복합성 피부 유형 : 예 2

피부미용사 : 고객님의 피부 유형은 복합성 피부에 해당합니다. 고객님의 피부는 지성과 건성이 모두 나타나는데 특히 코와 미간의 T존 위는 피지 분비가 많아 번들거리고, 이마와 볼은 건조하게 느껴지실 겁니다.

복합성 피부 유형 : 예 3

피부미용사 : 피부 상태의 균형을 맞추기 위한 관리를 하겠습니다. 지성과 건성 부위에 따른 두 가지 다른 종류의 마스크를 적용하겠습니다. 고객님의 T존 부위에는 각질 제거 마스크를, 이마와 뺨에는 수분 공급을 위한 마스크를 사용하겠습니다.

Oily skin type : Case 1

Esthetician : You know your skin type is oily. You have excess sebum, and your skin dehydrates. In other words, your skin lacks moisture. So your skin needs a balance of oil and water on your face. The last thing, I' m concerned about the comedones on your T-zone.

Oily skin type : Case 2

Esthetician : Your skin is greasy. There is excessive sebum on your skin and it lacks moisture due to dehydration. You need a balanced secretion of sebum and enough moisture. Finally I worry about black heads around your T-zone.

Oily skin type : Case 3

Esthetician : The main concerns are excess sebum and dehydrated skin. First, I would like to recommend extraction and deep cleansing for your first mask. Then I' d like to use an intensive hydrating mask.

Key words

balance : 균형
dehydration : 탈수, 건조
moisture : 수분
sebum : 유분
greasy : 지성인, 기름이 많이 묻은

지성 피부 유형 : 예 1

피부미용사 : 고객님의 피부 유형은 지성입니다. 피지는 많이 분비되는데 피부는 건조합니다. 즉 피부에 수분이 부족합니다. 그래서 고객님은 피부에 피지와 수분의 조화가 필요합니다. 마지막으로 T존 부위에 있는 여드름이 염려됩니다.

지성 피부 유형 : 예 2

피부미용사 : 고객님의 피부는 지성입니다. 피지는 과잉 분비되는데 탈수 때문에 수분은 부족합니다. 고객님은 충분한 수분과 피지 분비의 조화가 필요합니다. 특히 T존 부위의 블랙 헤드가 염려됩니다.

지성 피부 유형 : 예 3

피부미용사 : 아마도 고객님의 피부는 과잉 피지와 수분결핍이 문제일 것입니다. 우선 피지제거를 하고 첫 번째 마스크로 딥 클렌징 마스크를 추천하고 싶습니다. 그 다음으로는 수분 공급을 풍부히 하기위한 마스크를 사용하겠습니다.

Sensitive skin type : Case 1

Esthetician : The skin on your face is a higher temperature than the skin on your chest. Your cheeks get red. You have sensitive skin.

Customer : Yes, you' re right. What kind of treatment should I have?

Esthetician : I recommend a procedure that will adapt well to your condition. In my opinion, you should choose the product and procedure that will suit your condition and area of the skin the best. Depending on those factors, you may add or omit certain procedures.

Sensitive skin type : Case 2

Esthetician : Your cheeks and the sides of your nose have redness. This is caused by distended or broken capillaries. So we will avoid procedures and treatment that might irritate your skin. I am not going to use a brush machine, deep tissue massage, steam or hot towels.

Key words

adapt : 맞추다, 적응하다
avoid : 방지하다, 막다, 피하다
capilary : 모세혈관
distend : 팽창하다
irritate : 자극하다
omit : 빠뜨리다, 생략하다
opinion : 의견, 견해
sensitive skin : 예민성 피부

민감성 피부 유형 : 예 1

피부미용사 : 얼굴 부위의 피부 온도가 체온보다 높고 볼 부위가 붉은 것으로 보아 민감성 피부입니다.

고객 : 네, 맞아요. 어떤 종류의 관리를 받아야 할까요?

피부미용사 : 고객님의 피부 상태에 알맞은 피부 관리를 추천해드리겠습니다. 전문가적 견해로는 고객님은 고객님의 피부 상태나 부위에 가장 잘 맞는 제품이나 피부 관리를 선택해야합니다. 또 거기에 맞게 어떤 과정들은 더하거나 뺄 수도 있습니다.

민감성 피부 유형 : 예 2

피부미용사 : 고객님의 코 양쪽 부분과 뺨이 붉은데요. 이것은 팽창되거나 파열된 모세혈관이 원인으로 보입니다. 그래서 저희는 피부를 자극하는 과정이나 관리는 피하고, 브러시 기계나 심층 조직 마사지, 스팀과 뜨거운 타월 등은 사용하지 않겠습니다.

Customer Questionnaire

GENERAL QUESTIONS

Name			Today' s Date			
Address						
Phone						
Age	Under 21	21-30	31-40	41-50	51-60	Over 60
Occupation						

SKIN CARE QUESTIONS

1) What three things would you like to change about your skin?

( )

( )

( )

2) How much time are you willing to devote to your skin?

morning ( ) night ( )

3) Do you feel that your skin is sensitive? Yes ( ) No ( )

4) Please, check any of the following health challenges you have at the present time or have had in the past.

Eczema / Allergies / Cancer / Diabetes / Psoriasis / High Blood Pressure Varicose Veins / Heart Problems / Herpes Simplex Hormone Imbalance

Other ( )

5) List all medications that you take on a regular basis and reasons for usage:

( )

6) Have you taken the drug within the last year?

Yes ( ) No ( )

7) Have you had any cosmetic surgery in the area to be waxed at the any time in the past? Yes ( ) No ( )

(계속)

Key words

eczema : 습진

diabetes : 당뇨병

psoriasis : 건선, 버짐

high Blood Pressure : 고혈압

varicose Veins : 정맥류

heart Problems : 심장병

herpes Simplex : 단순 포진

hormone Imbalance : 호르몬 이상

Customer Questionnaire
8) If "yes", Where, and What type? ( )
9) Is your menstrual period due within the next week? Yes ( ) No ( )
10) Are you pregnant? Yes ( ) No ( )
11) Do you have regular exposure to the sun? Yes ( ) No ( )
12) Do you use any sun protectants for your skin? Yes ( ) No ( )
13) Have you used an Alpha Hydroxy, Beta Hydroxy Glycolic Acid or any other deep exfoliation product in the past week? Yes ( ) No ( )

5. Manual facial care

(1) Swedish massage

Esthetician : Let me give you a massage before applying your first mask. This massage stimulates your blood and lymph circulation. The pressure may be gentle. Let me use massage oil that will help calm and relax you. Please, let loose.

Customer : Ok!

Esthetician : I have finished the lymph massage. The next step is applying the first mask with steam. Before applying the mask, let me cover your eyes with pads and place tissues on your hair line. Now I' ll turn the steamer on. So can you feel the steam on your face?

Customer : Yeah. I can.

Esthetician : I will apply the first mask that helps slough off dead skin cells. Do you have any allergies I should be aware of or any sensitivity to the environment; temperature, moisture level, touch etc?

Customer : I' m not sure. I worry about ingredients from the sea.

Esthetician : I see. I will avoid the products that are made of ingredients from the sea. If you feel irritated from some products or uncomfortable, please let me know.

Customer : Thank you, Grace.

Key words

circulation : 순환
extract : 짜내다
ingredient : 성분, 원료, 재료
nourish : 자양분을 주다
secure : 안전한
slough off : 벗겨지다

5. 수기 얼굴 관리

> **Check point**
>
> 고객이 작성한 고객카드와 상담, 피부 분석을 통해 고객의 피부 상태를 정확히 판단하여 이에 따른 고객 맞춤 관리를 실행한다. 고객 피부에 맞는 각각의 시행 단계를 간단히 설명한다. 또한 관리 중에는 말을 할 수 없는 고객의 상태에 주의를 기울여야 한다.

(1) 스웨디쉬 마사지

피부미용사 : 첫 번째 마스크를 시행하기 전에 마사지를 하겠습니다. 혈액과 림프순환을 촉진하는 부드러운 압을 이용한 마사지입니다. 긴장을 풀 수 있도록 도와주는 오일을 사용하겠습니다. 편안히 계십시오.

고객 : 알겠습니다.

피부미용사 : 림프 마사지가 끝났습니다. 다음 단계는 스팀을 이용한 마스크입니다. 마스크를 시행하기 전에 아이 패드를 덮고 헤어라인에 티슈를 두르겠습니다. 이제 스팀 기계를 켰습니다. 고객님 얼굴에 스팀이 오는 것이 느껴지십니까?

고객 : 네, 느껴져요.

피부미용사 : 각질을 제거해주는 첫 번째 마스크를 시행하겠습니다. 고객님, 특정한 성분이나 환경, 예를 들어 온도나 습도, 압력, 접촉 등에 알레르기가 있습니까?

고객 : 확실하지 않지만 바다에 관련된 성분이 걱정되네요.

피부미용사 : 알겠습니다. 바다로부터 나오는 물질로 만들어진 제품을 피하도록 하겠습니다. 만약 제품으로부터 오는 자극을 느끼거나 불편하시다면 알려주세요.

고객 : 고마워요.

Esthetician : After removing the mask with a towel, I will apply the toner. It might be cold, please close your eyes. Now, I' m going to extract comedones and other blemishes on your nose. Are you going out tonight? After extraction your skin may be red for the remainder of today. So if you don' t want to do the extraction, we could skip it. It's up to you.

Customer : No, I don' t have a meeting tonight. It doesn' t matter. I want to remove my blemishes now.

Esthetician : The next step is a relaxing massage. The relaxing massage will relieve muscle tension and pain. I' ll use massage oil, it helps calm you down. Well, I finished the relaxing massage. How do you feel?

Customer : I feel great.

Esthetician : To secure your eyes and hair, let me place eye pads on your eyes and tissue on your hair line again. I will apply your second mask, the benefits include : tighter skin, increased hydration and better nourishment. Well, it' s time to remove your second mask. I' ll use a cold towel for calming and tightening your skin. With this, I have finished the facial procedure. Did you enjoy the service?

Customer : Yes, I did. Thank you. I feel refreshed.

Key words

benefit : 혜택, 이득
comedo(nes) : 여드름
open comdo : 흑면포 =black comdo
close comedo : 백면포 = white comedo
nourishment : 영양분, 자양분
remainder : 나머지
secure : 안전한, 확실한
skip : 건너뛰다
tension : 긴장상태

피부미용사 : 타월로 마스크를 지운 후 토너를 적용하겠습니다. 차가울 수 있습니다. 눈을 감아주세요. 이제 코에 있는 면포를 짜려고 하는데요. 오늘 저녁 약속 있으신가요? 면포를 짜낸 후에는 얼굴이 붉어질 수 있어요. 만약 짜내고 싶지 않으면 시행하지 않겠습니다. 고객님께서 결정해 주세요.

고객 : 아니오. 오늘은 약속이 없으니까 상관없어요. 저는 면포를 없애고 싶어요.

피부미용사 : 다음은 안정과 휴식을 주는 마사지입니다. 이번 단계에서는 편안함을 느끼고 근육의 긴장과 통증을 덜어줄 것입니다. 평온함을 느끼게 도와주는 마사지 오일을 사용하겠습니다. 자, 마사지가 끝났습니다. 기분이 어떠십니까? 괜찮으십니까?

고객 : 네, 좋아요.

피부미용사 : 다시 눈과 헤어를 보호하기 위해서 아이패드와 티슈를 눈과 헤어라인 위에 덮겠습니다. 두 번째 마스크를 바르겠습니다. 이 마스크는 탄력 있는 피부를 만들고 수분과 영양분을 공급해주며 피부를 진정시키는 효과가 있습니다. 이제 두 번째 마스크를 제거할 시간입니다. 피부의 진정과 탄력을 위해 차가운 타월을 사용하겠습니다. 이로써 얼굴관리가 모두 끝났습니다. 이 서비스에 대해 만족하셨나요?

고객 : 네, 고마워요. 느낌이 산뜻하네요.

(2) Korean facial care

Esthetician : I will introduce Korean massage briefly. The process begins with washing the face, and then a skin analysis, a massage which opens the path of blood and 'Ki' , a Korean word referring to the 'body' s energy, in your body, effectively relaxing five viscera, the six entrails and your skin and muscles. You will feel a little stronger pressure from the Korean facial skin care than from the basic massage.

Customer : That is good.

Esthetician : After your facial massage, how would you like to improve your skin?

Customer : I' m not sure, but I hope to make my face appear smaller.

Esthetician : I see. In my opinion, right now your skin is a little dry and your face is a little swollen. If you get a Korean massage, we will work to your face look slender and it will also be helpful in moisturizing your dry skin. How do you treat your skin now?

Customer : I don' t use anything special, just basic care products at home.

Esthetician : Ok, I see. After the Korean massage, to get enough moisturizing, we will apply a special nutrition product on your skin. Finally, we will conduct a special mask which will set quickly and heat your skin after applying it. Do you have claustrophobia or are you sensitive about heat on your skin? If not, after the massage, your skin will become moist and your face will look slender.

Customer : That sounds great. Thank you.

Key words

agoraphobia : 광장공포증
blood : 혈액, 피
claustrophobia : 폐쇄공포증
five viscera : 오장, liver, heart, spleen, lungs, kidney
moisturizing : 수분효과
muscle : 근육
nutrition : 영양
opinion : 의견, 견해
path : 통로, 길
quickly : 빨리
relax : 이완, 안심
six entrails : 육부, stomach, large intestine, small intestine, gall bladder, urinary bladder, triple burner
slender : 날씬한, 호리호리한

(2) 한국형 얼굴 관리

피부미용사 : 한국형마사지 서비스에 대해 간단하게 소개하겠습니다. 관리 진행은 세안, 피부분석, 기(氣)와 혈(血)이 순행하는 길이 있는데, 여기에 에너지인 기가 흐르는 통로를 열어주는 마사지를 한국형 마사지라 하며 인체의 오장육부와 피부, 근육까지 풀어주는 마사지입니다. 한국형 마사지는 매뉴얼 마사지보다는 좀 강하게 압력을 느끼실 수 있는 마사지입니다.

고객 : 좋은데요.

피부미용사 : 피부 관리 후 고객님의 피부가 어떻게 개선되었으면 좋겠습니까?

고객 : 정확히 모르겠어요. 하지만 얼굴이 좀 작아졌으면 좋겠다는 바람이 있어요.

피부미용사 : 알겠습니다. 제 생각에는 고객님 피부 상태는 조금 건조하시고 얼굴이 좀 부었어요. 한국형 피부 관리를 받으시면 얼굴 모양도 축소되고 건조한 피부의 보습에도 도움이 될 거예요. 지금은 어떻게 피부를 관리하시나요?

고객 : 저는 특별한 거는 사용하지 않고 그냥 집에서 일반적인 제품만 사용하고 있는데요.

피부미용사 : 네. 그렇다면 한국형 마사지 후 보습을 충분히 하실 수 있도록 고객님 피부에 특별 영양 앰플을 발라드릴게요. 마지막으로 그 위에 특수 팩을 해드릴 텐데 그것을 도포하면 곧 굳어지면서 피부를 따뜻하게 열이 나게 해줄 겁니다. 혹시 폐쇄 공포증이 있으시거나 열에 민감하지는 않으신지요? 괜찮으시다면 마스크 제거 후에는 피부가 촉촉해지고 얼굴도 작아진 느낌을 받으실 수 있을 것입니다.

고객 : 좋을 것 같네요. 고마워요.

(3) Lymph facial care

Key words

agitated : 뒤흔들다
circulation : 순환, 유통
collagen : 콜라겐
discharge : 빼내다, 흘러 나오게 하다(고름), = drain
improve : 개선하다
lymph : 림프(액), 진물
moisture : 습기, 수분
metabolism : 물질대사
edema : 부종
proceed : 계속해 나가다
pimple : 여드름, = acne, comedo
elasticity : 신축성, 탄력성
nutrition : 영양
promote : 촉진시키다
retain : 보유하다
wastes : 노폐물
secretions : 분비, 분비물
swollen : 부은 상태의
removal : 제거
technical : 기술적, 전문적인
wrinkles : 주름, a crow's foot /lines/furrows/ fine lines 주름살
diagnosis : 진단
be prone to : .. 의 경향이 있는, 하기 쉬운
allergy : 알레르기
be a good fit 잘 맞다
go down : 가라앉다
capillary vessels. 모세혈관
dilate : 확장시키다
dilatation : 확장

Esthetician : Let me briefly tell you about our lymph massage. As a manual skill which is created by Emil Vodder, the lymph massage promotes the circulation of lymphs and helps the metabolism of cells, and also helps to discharge wastes from the body, it improves nutrition metabolism and problems such as edema of the body. Before proceeding with your treatment, we will wash your face and complete a skin analysis. After that, you can get a lymph massage with the option of having a gel type collagen mask. This will improve your skin' s elasticity and moisture, and even ease your wrinkles. A lymph massage is soft and comfortable. But it will specially help the sensitive skin that is prone to allergy or pimples and the skin with dilatation of capillary vessels.

Customer : My skin is very sensitive, so that seems like a good fit.

Esthetician : After your facial massage, how would you like to improve your skin?

Customer : I' m not sure. But I want to improve my sensitive skin. It always seems to be agitated and turns red often. And sometimes my face becomes easily swollen.

Esthetician : I see. In my opinion, there needs to be a technical diagnosis here, if you get a lymph massage, your swollen face will go down and your skin will look and feel less agitated. How are you managing your skin at home?

(3) 림프 얼굴 관리

피부미용사 : 림프마사지 서비스에 대해 간단하게 소개하겠습니다. 에밀 보더(Emil Vodder) 박사에 의해 창안된 수기법으로 림프 순환을 촉진시켜 세포의 대사물질 및 노폐물의 체외배출을 도와주는 기법으로 조직의 영양대사를 원활히 하고 신체의 부종과 통증 미용 이상증과 같은 문제점 개선을 도와주는 방법입니다.
관리 진행은 세안, 피부분석, 후 림프 마사지를 받으실 수 있으며 마사지 후에는 젤 형태의 콜라겐 마스크 받으시면 마스크로부터 피부의 탄력 증진 수분공급 주름 완화 효과를 받으실 수 있습니다.
림프마사지는 부드러우면서도 편안하게 받으실 수 있고 알레르기나 여드름, 모세혈관 확장피부 등 자극에 민감한 피부에 도움을 받으실 수 있습니다.

고객 : 제 피부가 무척 예민하니까 도움이 되겠군요.

피부미용사 : 피부 관리 후 고객님의 피부가 어떻게 개선되었으면 좋겠습니까?

고객 : 정확히 잘 모르겠어요. 하지만 예민해서 붉어지는 증상이 좀 개선되었으면 좋겠어요. 그리고 제 얼굴이 잘 부어요.

피부미용사 : 알겠습니다. 제 생각에는 림프 관리를 받으시면 우선은 부은 얼굴이 가라앉고 붉어지는 증세가 서서히 좋아질 것입니다. 집에서 특별히 하시는 관리는 있으신가요?

Customer : I just use general care products at home and don' t use a special treatment plan. When my face becomes red, sometimes, I use a cold aloe gel.

Esthetician : OK, that' s helpful. So after the lymph massage, to provide your skin with enough moisture, I will apply a gel pack type of collagen so you will feel a little cool. After the removal of the mask, your skin will retain more moisture and its elasticity will have improved.

Customer : That will be great. Thank you.

고객 : 아니요 저는 그냥 집에서 일반적인 제품만 사용하고 특별히 하는 관리는 없어요. 얼굴이 붉게 달아오를 경우 가끔 알로에 젤을 냉장고에 시원하게 해두었다가 팩으로 사용하기도 해요.

피부미용사 : 네 좋아요. 그렇다면 림프마사지 후 보습을 충분히 하실 수 있도록 콜라겐 젤 타입의 팩으로 관리를 해드리겠습니다. 좀 시원한 느낌을 받으실 텐데 괜찮으시다면 마스크 제거 후에는 피부가 촉촉해지고 탄력이 증진된 느낌을 받으실 수 있을 것입니다.

고객 : 좋을 것 같네요. 고마워요.

6. Finishing touch

Esthetician : We have completed all of the facial skin care procedures. Please close your eyes. Let me spray the toner again. I will apply eye and lip cream first, and then I' ll apply moisturizer, day cream and sunscreen to protect your skin from the exterior environment.

Customer : Thanks.

Esthetician : May I apply the finishing touch? The finishing touch will be minimized makeup. Specifically I will minimize the makeup on your eyebrows and lips. However this will not treat your skin, only add to its aesthetic beauty.

Customer : I see.

Esthetician : Ok! Please, have a seat here and close your eyes. It takes only a few minutes. I will use an angle brush and brown shadow. It makes your brows look natural. Finally I will apply lip gloss. I think this color fits on your lips well. Would you like to try this one?

Customer : Yes, that' s my favorite color.

Esthetician : What do you think? If you want to correct anything, let me know.

Customer : It' s nice, no problem.

Key words

exterior environment : 외부 환경

eyebrow : 눈썹

minimize : 최소화하다, 축소하다

6. 마무리 관리

피부미용사 : 이제 얼굴 관리를 모두 마쳤습니다. 눈을 감아주세요. 다시 토너를 뿌리겠습니다. 먼저 아이크림과 립크림을 바르고, 로션과 데이크림, 그리고 외부환경으로부터 피부를 보호하기 위한 자외선 차단제를 바르겠습니다.

고객 : 네.

피부미용사 : 제가 마무리 관리를 해도 될까요? 마무리(Finishing touch)란 최소한의 메이크업입니다. 정확하게 말하면, 눈썹과 입술 화장을 하는 것입니다. 만약 고객님이 괜찮으시다면 마무리 관리를 하겠습니다. 하지만 이 단계는 피부 관리가 아니라 다만 고객님의 외모가 좋게 보이기 위한 것입니다.

고객 : 네, 해주세요.

피부미용사 : 좋아요. 눈을 감고 앉아주시겠습니까? 몇 분 정도면 됩니다. 앵글 브러시와 갈색의 아이섀도를 사용하는데 이것은 눈썹을 자연스럽게 만들 수 있습니다. 마지막으로 립글로스를 바르겠습니다. 제 생각에는 이 색상이 고객님의 입술에 어울릴 것 같은데요. 이 색상이 마음에 드십니까?

고객 : 네. 제가 좋아하는 색상이네요.

피부미용사 : 어떠세요? 마음에 드십니까? 수정하시고 싶으면 말씀해 주십시오.

고객 : 좋습니다.

CHAPTER 03

Body care

1. Foot Reflexology

Esthetician : Hi, Mrs. Crystal, I am Annie, a therapist here. Today I' ll give you a variety of massages. At first, I' ll introduce you to our reflexology service in a simple way. Reflexology is a type of alternative medicine that has been passed down orally for five thousand years. In reflexology, a sole is thought of as a miniature human body, the stimulation of certain areas of the sole corresponds with internal organs. This helps each internal organ to function smoothly. We are performing the reflexology service, based on those fundamentals.

Customer : OK, now my feet are really sore and tired from a long time shopping, I think it will be good for me.

Esthetician : Of course. First, I' ll let your feet soak in warm water. A warm foot bath improves the circulation of your body' s fluids and is great way to rejuvenate your body in a short time. Please soak your feet in the warm water. How' s the temperature?

Customer : I thought it was a little bit hot, but it' s ok now.

Key words

- **alternative medicine** : 대체의학, 건강요법
- **hurt** : 다치게 하다, 아프게 하다
- **miniature** : 축소모형
- **press** : 누르다
- **reflexology** : 반사학
- **seldom** : 좀처럼, 드물게
- **stick** : 봉, 막대기, 지팡이
- **stimulation** : 촉진, 자극, 흥분
- **supporting** : 지탱하는, 뒷받침하는
- **whole** : 전체의, 모든, 온전한

제3장 피부관리

1. 발반사 요법

피부미용사 : 안녕하세요, 크리스탈님. 오늘 고객님의 관리를 진행할 테라피스트 애니입니다. 먼저 발마사지 서비스에 대해 간단하게 소개하겠습니다.
발반사요법은 무려 5,000년 전부터 구전되어 내려오던 건강요법입니다. 발바닥을 인체의 축소판이라고 생각하여 발의 특정 부위를 자극하면 반사에 의해 해당 오장육부의 기능을 원활히 하도록 돕는다는 원리에 따라 테라피를 시행합니다.

고객 : 아, 제가 쇼핑을 오래 해서 발바닥이 아프고 몸도 피곤한데 도움이 되겠네요.

피부미용사 : 물론입니다. 먼저 따뜻한 물로 족욕을 하겠습니다. 족욕은 인체 순환을 촉진시켜 피곤한 몸을 빠른 시간에 회복시키는 데 매우 좋습니다. 이 물에 발을 담그십시오. 온도는 괜찮으십니까?

고객 : 네, 좀 뜨겁다 싶었는데 금세 적응이 되는데요.

Key words

Body Form of female
① bag of bones
② big-boned
③ boney
④ chubby
⑤ overweight
⑥ plump
⑦ skinny
⑧ slender
⑨ slim
⑩ thin

Body form of male
① beefy
② compact
③ heavy set
④ muscular
⑤ robust

Esthetician : At first, now I' ll remove the dead skin from your feet before the reflexology massage. Please lie on your belly and relax and make yourself comfortable.

We have two sorts of reflexology massages : one stimulates using a wooden stick, while the other stimulates using hands. Which do you prefer?

Customer : A wooden stick? Won' t that hurt?

Esthetician : No, it seldom hurts, even though I use the stick. But you might feel a little pain when I touch a certain area that may be related to an unhealthy organ.

Customer : Please use the stick.

Esthetician : How about this area? Does it hurt when I press here?

Customer : A little bit.

Esthetician : How about now?

Customer : No problem. I' m OK now.

Esthetician : I' ll apply a foot moisturizer as a finish. The massage is done. Our feet play a vital role in supporting our whole body, so please care for your feet and keep them clean.

Customer : I see. Thank you.

피부미용사 : 이제 간단히 각질을 제거한 후 마사지를 하겠습니다. 침대 위에 엎드려 편히 누워 주십시오.
발반사요법에는 봉을 이용하는 방법과 손으로만 자극하는 방법이 있습니다. 어느 방법이 좋으십니까?

고객 : 그 봉을 이용해서 마사지 하는 건가요? 아프지 않나요?

피부미용사 : 네, 이 봉으로 마사지를 해 자극을 줍니다만 아프지는 않습니다. 단 몸에 이상이 있는 반사구 부위는 통증이 조금 있을 수 있습니다.

고객 : 그럼 봉을 이용해서 해주세요.

피부미용사 : 이 부분은 어떻습니까? 누르면 아픈가요?

고객 : 네, 약간.

피부미용사 : 지금은 어떠십니까?

고객 : 괜찮습니다.

피부미용사 : 마무리 관리로 발 전용 보습크림을 발라드리겠습니다.
이제 관리가 다 끝났습니다. 발은 우리 인체를 지탱하는 역할을 하기 때문에 평소에도 관심을 갖고 청결히 관리해 주시면 좋습니다.

고객 : 알겠습니다, 고맙습니다.

2. Leg Massage

Esthetician : Hi, Mrs. Smith. The leg massage will last for 40 min. As you know, the leg is a very important part of our body having three- fourths of the body’ s muscles. It is, thus, apt to have much stress and stagnant blood flow. The leg massage will improve your blood circulation and ease your muscles.

Customer : OK.

Esthetician : I’ ll massage the back side of your leg and then the front side of it. Please lie on your belly and stretch out your knees.

Customer : I see, you mean, like this?

Esthetician : That’ s right, thank you. Does any part of your body feel uncomfortable?

Customer : Recently, my calf muscles have been continuously tight.

Esthetician : Yes, I can feel that. I’ ll do my best to relax and loosen the muscles in that area.

Customer : Thank you.

Esthetician : I finished the back side massage of your leg. Now for the front side massage of it, please lie on your back.

Customer : OK.

Esthetician : I’ ll stretch your legs. Breathe in and out smoothly. If you feel anything uncomfortably, please tell me.

Key words

breathe : 호흡하다, 숨을 쉬다
effleurage : 쓰다듬기
feel : 느끼다
important : 중요한
loosen : 느슨하게 하다, 풀다
motion : 동작
smoothly : 부드럽게, 순조롭게

2. 다리 마사지

피부미용사 : 안녕하세요, 고객님. 고객님이 원하시는 다리마사지는 약 40분 정도 진행될 예정입니다. 아시겠지만 다리는 인체 근육의 4분의 3이 모여 있다고 할 만큼 중요한 신체 부위지요. 그만큼 무리도 많고 쉽게 수분의 정체가 생기기도 합니다. 오늘은 다리마사지를 통해 혈액순환을 증진시키고 근육의 피로를 풀어드리겠습니다.

고객 : 네.

피부미용사 : 먼저 다리의 뒷면 관리를 한 후에 앞면 관리를 하겠습니다. 다리를 쭉 펴고 엎드려 누워주십시오.

고객 : 네, 알겠습니다. 이렇게 누우면 되나요?

피부미용사 : 네, 좋습니다. 고맙습니다. 혹시 특별히 불편한 점이 있으십니까?

고객 : 제가 요즘 종아리 부분이 계속 당기는 느낌을 받습니다.

피부미용사 : 네, 약간 뭉쳐있네요. 그 부분을 조금 더 신경 써서 풀어드리겠습니다.

고객 : 네.

피부미용사 : 이제 뒷면 관리가 끝났습니다. 앞면 관리를 시행하겠습니다. 앞으로 돌아누워 주십시오.

고객 : 네.

피부미용사 : 다리 스트레칭을 하겠습니다. 고객님은 힘을 빼고 편안히 누워 계시면 됩니다. 숨을 천천히 들이마셨다 내쉬십시오. 언제든 불편하시면 말씀해 주십시오.

Customer : Sure.

Esthetician : I' ll finish this massage with an effleurage motion. How was it? Do you feel the difference?

Customer : Yes, I feel fresh on my legs.

Esthetician : Great!

3. Hand Massage

Esthetician : How are you today, Mrs. Jolly. I am your hand massage therapist Jennie today. A hand massage can help the circulation in your hands, relieve their stress and, above all, make your hands moist and beautiful. At first, I' ll wipe your hands with a warm towel and get rid of the old skin using the hand scrub cream.

Customer : Do you also need to remove the dead skin?

Esthetician : Of course, If your hands have dead skin cell, your hands won' t feel soft but rough. Also, even if you apply quality hand cream, you won' t benefit from it, since there will be little absorption. When you exfoliate them with hand scrub cream, you can have much cleaner and softer hands.

Customer : Oh, I see. I thought just washing them carefully was enough.

Esthetician : I' ll give you a hand massage with aroma oil. Do you have a favorite fragrance?

Key words

amazing : 놀라운
benefit : 혜택, 이득
blend : 섞다, 혼합하다
carefully : 조심스럽게, 주의해서
enough : 충분히, 충분한
exfoliate : 벗겨내다, 박리하다
favorite : 마음에 드는, 좋아하는
fruits : 과일
relieve : 완화시키다, 안도하게 하다
rid : 제거하다, 자유롭게 하다
scent : 향기 향내
smell : 냄새가 나다
start : 시작하다, 출발하다
wipe : 닦다

고객 : 네.

피부미용사 : 이제 에플라지 동작으로 마무리 하겠습니다. 어떠십니까? 다리의 피로가 많이 풀린 것 같습니까?

고객 : 네. 아주 개운합니다.

피부미용사 : 만족하셨다니 다행입니다.

3. 손 마사지

피부미용사 : 안녕하세요, 고객님. 오늘 고객님의 핸드마사지를 진행할 테라피스트 제니입니다. 핸드마사지는 순환 개선에 도움을 주어 피로를 풀어주고 무엇보다 손을 촉촉하고 아름답게 만들어 주는 효과가 있습니다. 먼저 온습포로 닦아낸 다음 스크럽을 이용해 각질 제거를 하겠습니다.

고객 : 손도 각질 제거가 필요한가요?

피부미용사 : 물론입니다. 손에 각질이 많으면 손이 부드럽지 못 하고 거친 느낌이 납니다. 또한 피부 흡수율이 떨어지므로 아무리 좋은 크림을 발라도 효과를 기대하기 힘들지요. 이렇게 바디용 스크럽으로 각질을 제거하면 훨씬 맑고 부드러운 손으로 만드실 수 있습니다.

고객 : 그렇군요. 저는 잘 씻기만 하면 되는 줄 알았어요.

피부미용사 : 이제 아로마오일을 이용해 핸드마사지를 하겠습니다. 블렌딩하기 전에 좋아하는 향이 있으신지요?

Customer : I like the refreshing smell of citrus fruits like mandarin orange. Would it be good for my hands?

Esthetician : Of course. I' ll blend jojoba oil with neroli oil, and massage with it. Do you want to get the scent?

Customer : Sure. That would be amazing!

Esthetician : I' ll start the hand massage now. Please let me know if you feel uncomfortable at any point.

Customer : I see, thank you.

4. Back Massage

Esthetician : Hi, Mrs. Fogel!

Customer : Hi, I' d like to have a back massage.

Esthetician : OK, I' ll prepare it. Please change into this gown in the fitting room. Wear the gown over the disposable underwear.

Customer : Ok.

Esthetician : After you' ve changed, please come this way. Please lie on your belly and relax. Are you cold? I heated the room already.

Customer : I' d like it to be warmer, if you can.

Esthetician : Sure, wait a moment and I' ll adjust the temperature. Would you come upward a little?

Customer : OK.

Key words

- **adjust** : 조정하다, 적응하다
- **area** : 지역, 부분
- **begin** : 시작하다
- **belly** : 배
- **ease** : 편안함, 안락함
- **focus** : 집중하다, 맞추다
- **gentle** : 온화한, 조심스러운
- **heat** : 열, 열기
- **hot compress** : 온습포
- **pain** : 통증
- **shoulder** : 어깨
- **upward** : 위쪽을 향한

고객 : 전 감귤냄새같이 상큼한 향을 좋아하는데 손에도 좋은가요?

피부미용사 : 네, 그럼 호호바에 네롤리오일을 함께 블렌딩하여 마사지를 하겠습니다. 향을 한 번 맡아 보시겠습니까?

고객 : 네, 아주 좋습니다.

피부미용사 : 그럼 핸드마사지를 시작하겠습니다. 마사지 중간이라도 불편하신 점이 있으시면 말씀해 주십시오.

고객 : 네, 고맙습니다.

4. 등 마사지

피부미용사 : 안녕하십니까, 고객님?

고객 : 안녕하세요? 등 관리를 좀 받고 싶어서 왔는데요.

피부미용사 : 네, 준비하겠습니다. 우선 탈의실에서 가운을 갈아입고 나오십시오. 상의는 완전 탈의하시고 라커 안에 준비되어 있는 일회용 팬티를 착용하신 후 가운을 입으시면 됩니다.

고객 : 알겠습니다.

피부미용사 : 다 갈아 입으셨으면 이쪽 관리실로 오십시오. 침대에 엎드려 편히 누우십시오. 온도를 높여 놓았는데 혹시 춥지는 않으십니까?

고객 : 약간 더 따뜻했으면 좋겠네요.

피부미용사 : 네, 알겠습니다. 온도를 조금 더 높였으니 잠시만 기다려 주십시오. 죄송하지만, 약간만 위쪽으로 조금 더 올라와 주시겠습니까?

고객 : 알겠습니다.

Key words

Body Form of female

① bag of bones
② big-boned
③ boney
④ chubby
⑤ overweight
⑥ plump
⑦ skinny
⑧ slender
⑨ slim
⑩ thin

Body form of male

① beefy
② compact
③ heavy set
④ muscular
⑤ robust

Esthetician : I' ll start to cleanse your back with a hot compress before we begin. Let me know if you feel uncomfortable at any time.

Customer : Okay. I like gentle pressure. So please do it softly.

Esthetician : I' ll do it softly with an aroma oil, to ease your muscles. Please make yourself at home. Am I applying to much pressure?

Customer : No, that' s good.

Esthetician : Your shoulder muscle is very tight. I' ll focus on this area. During the massage, please tell me if you feel any pain. Are you OK?

Customer : I feel a little pain, but I think it is refreshing.

5. Abdominal Massage

Esthetician : I' ll introduce our abdominal massage therapy today. After cleansing the area we' ll scrub the old skin from your belly and an apply anti cellulite cream.

Customer : If so, can I be wasp-waisted ?

Esthetician : Sure, I' ll do my best.

Customer : Thanks, I really appreciate it.

Esthetician : I' ll apply a hot compress to it.

Customer : I feel great now, because the compress is hot.

Key words

abdomen : 복부
cellulite : 셀룰라이트, 귤껍질모양의 피부
introduce : 소개하다
nerve : 신경
request : 요구하다

피부미용사 : 네, 좋습니다. 이제 간단히 온습포로 클렌징 한 후 마사지를 시작하겠습니다. 불편하신 점이 있으면 언제든 말씀해 주십시오.

고객 : 네, 제가 원래 부드러운 마사지를 좋아하는 편이라 너무 센 마사지는 피해주세요.

피부미용사 : 네, 알겠습니다. 아로마오일을 이용해 부드럽게 근육을 풀어주는 마사지를 해드리겠습니다. 긴장을 풀고 편안하게 계십시오. 압은 어떠십니까?

고객 : 네, 좋습니다.

피부미용사 : 어깨 근육이 많이 뭉치셨네요. 이 부분을 집중적으로 마사지하겠습니다. 아프면 말씀해 주세요. 지금 괜찮으십니까?

고객 : 네, 약간 통증은 있지만 시원합니다.

5. 복부 마사지

피부미용사 : 오늘 진행할 복부마사지 프로그램에 대해 안내해 드리겠습니다. 먼저 클렌징을 한 후 스크럽을 이용해 노화각질을 제거할 것입니다. 그 다음에 아로마오일을 이용해 복부마사지를 한 후 셀룰라이트 분해크림을 발라드릴 것입니다.

고객 : 아, 그렇게만 하면 제 허리도 개미허리가 될 수 있나요?

피부미용사 : 물론이죠, 그렇게 되실 수 있도록 최선을 다하겠습니다.

고객 : 네, 기대됩니다.

피부미용사 : 먼저 온습포로 클렌징을 하겠습니다. 따뜻합니다.

고객 : 네, 습포가 따뜻하니까 몸이 확 풀리는 것 같습니다.

Esthetician : Oh, that' s it. This scrub is really useful for getting rid of the dead skin cells and wastes.

Customer : Great.

Esthetician : Now I' ll massage your abdomen with sweet almond oil blended with fennel. I' ll apply more pressure than other massages because there are a lot of nerves and lymph nodes inside the abdomen. But you cannot feel much pain because there is much fat in that area. I' ll proceed with the massage along the flow of the organs. How is the pressure?

Customer : It' s OK.

Esthetician : Lastly, I' ll apply anti cellulite cream now. As a gel type cream, it absorbs very well. The area will be a little warm for about an hour. If you use exercise and dietary manipulation, you will see better results.

Customer : I see. I' ll try harder this time.

Esthetician : We' ll do our best to fulfill your requests, too. Thank you.

Key words

pressure : 압력
wastes : 폐수, 불순물, 폐기물
lymph node : 림프절
absorb : 흡수
flow : 흐르다
dietary manipulation : 식이 조작
request : 요청, 요구

피부미용사 : 네, 고맙습니다. 이 스크럽은 바디 전용으로 복부에 쌓여 있는 각질과 노폐물 등을 제거하는 데 아주 유용합니다.

고객 : 네.

피부미용사 : 이제 스윗아몬드오일에 휀넬을 블렌딩한 오일로 복부를 마사지하겠습니다. 일반적으로 복부는 신경이나 림프가 안쪽으로 있기 때문에 다른 마사지보다 조금 더 압력을 줍니다만 지방이 많아서 아프지는 않습니다. 마사지의 방향은 장기의 흐름에 따라 진행합니다. 압력은 어떠십니까?

고객 : 괜찮습니다.

피부미용사 : 마지막으로 셀룰라이트 분해크림을 발라드리겠습니다. 젤 타입으로 흡수력이 매우 좋습니다. 한 시간 정도 따뜻한 느낌이 날 겁니다. 식이요법과 운동을 병행하시면 더욱 좋은 결과를 보실 수 있을 겁니다.

고객 : 네, 이번 기회에 꼭 효과를 보도록 노력해야죠.

피부미용사 : 저희도 최선을 다하겠습니다. 고맙습니다.

SPA Expression

Preparation

[Case I]

① Good morning. I am your therapist Ann.
안녕하십니까? 테라피스트 앤 입니다.

② Are you Mrs. Crystal?
크리스탈 고객님, 맞으시죠?

③ I will show you to the changing room.
제가 탈의실로 안내해 드리겠습니다.

④ Please change into your robe, I will be back soon.
여기서 가운으로 갈아입으십시오. 잠시 후에 돌아오겠습니다.

⑤ Have you showered? Will you shower?
샤워는 하셨습니까? 샤워를 하시겠습니까?

⑥ I will show you to the Sauna.
사우나로 안내해 드리겠습니다.

⑦ You can see the ocean while you are in the Sauna during the morning and afternoon hours.
저희 사우나는 아침과 낮에 바다를 보면서 이용하실 수 있습니다.

⑧ Cleansing soap is a gel type and you can use it with water.
세안 비누는 젤 타입으로 물에 섞어 사용하실 수 있습니다.

⑨ Shower towel and a regular towel are here.
샤워 타월과 타월은 여기 있습니다.

⑩ We have a combined shampoo and conditioner, toothpaste and soap.
샴푸와 린스 겸용이 있고 치약과 비누가 준비되어 있습니다.

⑪ The shower room and Sauna are inside, over looking the ocean.
샤워장과 사우나는 안쪽의 바다가 보이는 곳에 있습니다.

⑫ If you would like to use the swimming pool, please shower inside first.
수영장 이용을 원하시면 안쪽에서 샤워를 하십시오.

⑬ If you need anything, just let me know anytime.
더 필요한 사항이 있으시면 언제든지 제게 말씀해 주십시오.

⑭ After you change into your robe, I will escort you to the massage room.
가운을 입으시면 제가 마사지 룸으로 안내해 드리겠습니다.

⑮ You can change here.
여기에서 옷을 갈아입으십시오.

⑯ Please, turn your key to the left and pull it.
키는 왼쪽으로 돌려서 잡아당기면 됩니다.

[Case II]

① Please take off your robe.
가운을 벗어 제게 주십시오.

② You can wear these underwear anyway.
일회용 팬티는 앞과 뒤가 같습니다.

③ Please lie here with your face down. Are you comfortable?
얼굴을 아래로 향하게 엎드리십시오. 편안하십니까?

④ I will dim the light a bit.
조명등을 조금 낮추어 드리겠습니다.

⑤ Please relax your body.
몸에 힘을 빼주십시오.

⑥ How is the water temperature?
물의 온도는 괜찮습니까?

⑦ Would you like to have a strong or weak massage?
마사지는 강하게 해드릴까요, 약하게 해드릴까요?

⑧ Are you alright with this pressure?
이 정도의 압력이 괜찮으십니까?

⑨ If you feel pain or discomfort, please tell me.
마사지를 받으면서 아프시거나 불편하시면 저에게 말씀해 주십시오.

⑩ Do you have any special part of your body that you want to treat?
오늘 특별히 마사지 받고 싶은 부위가 있으십니까?

⑪ Your shoulder is hard. Did you sleep well last night?
어깨의 근육이 뭉쳤네요, 지난밤에 잠은 잘 주무셨습니까?

⑫ Are you cold or hot?
추우십니까? 더우십니까?

⑬ Is the music too loud for you?
음악 소리가 너무 크지는 않습니까?

⑭ Please lie on your back.
얼굴을 위로 하고 누우십시오. (돌아누우십시오)

⑮ This mask will help your blood circulation and make you warm.
이 마스크는 혈액 순환을 시켜 주고 따뜻하게 해줍니다.

⑯ This mask is a bit cold(or hot).
이 마스크는 조금 차갑습니다. (조금 뜨겁습니다.)

⑰ This is a water mask. It is a nutrition mask.
보습 마스크입니다. 영양 마스크입니다.

⑱ I will use a paraffin mask on your foot.
발에 파라핀 마스크를 해드리겠습니다.

⑲ I will finish in about 5minutes.
5분 후에 마무리 해드리겠습니다.

⑳ We are almost finished. Is it alright?
마사지 시간이 조금 지났는데 괜찮으십니까?

㉑ The massage is finished. Please sit down.
마사지가 끝났습니다. 앉아 주십시오.

㉒ Please wait a minute. I will massage your shoulder.
잠시만 기다리십시오. 어깨를 만져 드리겠습니다.

㉓ Was any part of the massage uncomfortable?
마사지를 받으시면서 불편한 곳은 없으셨습니까?

㉔ Your locker key is in your robe pocket.
라커 키는 가운 주머니 안에 있습니다.

㉕ If you want to take a shower, please do it quickly and use warm water.
샤워하고 싶으시면 따뜻한 물로 간단히 해주십시오.

㉖ If you take a shower with warm water before you go to bed, you will get a better night sleep.
오늘밤 잠자기 전에 어깨를 따뜻한 물로 다시 한 번 샤워해주시면 잠이 잘 올 것입니다.

㉗ Did you have a good time?
편안한 시간이 되셨습니까?

㉘ I recommend you have the back massage next time.
다음 기회엔 등 마사지를 받아보십시오

㉙ Your locker is this way.
라커는 이쪽입니다.

㉚ Please throw away your underwear.
입고 계신 팬티는 여기에 버리시면 됩니다.

㉛ The comb is in a sterilizer.
머리빗은 소독기 안에 있습니다.

㉜ We hope to see you again.
다음에 또 뵙겠습니다.

[Case III]

① Good morning / afternoon / evening. May I help you?
안녕하십니까? 무엇을 도와 드릴까요?

② Would you like to try the massage?
마사지를 한 번 받아 보시겠습니까?

③ Here is the list of massage.
여기 마사지에 대해 자세한 메뉴 리스트가 있습니다.

④ Do you prefer one over another?
특별히 원하시는 마사지가 있습니까?

⑤ Do you feel stress, tension or tightness in a specific part of your body?
특정한 부위에 스트레스, 긴장감, 혹은 불편하게 느끼는 곳은 없습니까?

⑥ If you' d like my recommendation, just ask!
저희가 추천해 드리겠습니다.

⑦ You should book 3 hours in advance.
마사지는 3시간 전에 예약하시면 됩니다.

⑧ 10 minutes later, you can have massage.
10분 후에 마사지 받으실 수 있습니다.

⑨ Unfortunately, we are all booked up at that time.
죄송합니다. 그 시간은 이미 예약이 차있습니다.

⑩ Yes, you can have it now.
네, 지금도 가능합니다.

⑪ I will make a reservation under Susan Smith, room number 1109 for the scalp massage at 11 am.
오전 11시 두피마사지로 방 번호 1109, 성함은 수잔님으로 예약해 드리겠습니다.

⑫ Would you like to use the Sauna before the massage?
마사지 전에 사우나를 이용하시겠습니까?

⑬ If you want to use the Sauna, please come earlier than your reservation time.
사우나를 이용하시려면 예약시간 전에 시간을 맞추어 이곳으로 오시면 됩니다.

⑭ Thank you! See you at 2 pm. Ms. Susan Smith.
수잔 고객님, 오후 2시에 뵙겠습니다.

CHAPTER 04

Makeup Practice

1. Initial Consultation

Makeup artist : May I help you?

Customer : Hello, I’ ve actually never been here before.

Makeup artist : Welcome to Trendy Salon! I’ m Grace the manager. What can we do for you?

Customer : I would like to make an appointment for makeup.

Makeup artist : What kind of makeup do you want? Is there any particular makeup you have in mind? I can recommend a professional makeup artist according to the style you want.

Customer : What kind of makeups do you offer?

Makeup artist : We have a variety of T.P.O. makeup such as beauty makeup, photo makeup, wedding makeup and stage makeup.

Customer : I’ m afraid I don’ t know them well. How do they differ from one another?

Makeup artist : Beauty makeup is a regular makeup offered by a makeup artist to improve your look. This style is mainly for a casual visit, a personal interview, meeting with customers, a gathering, and an interview with a prospective

Key words

according to : ...에 따르면, ...에 따라
actually : 실제로, 정말로
bride : 신부
bridegroom : 신랑
circumstance : 상황, 환경
departure : 출발
deposit : 예약금
exclusively : 독점적으로, 오로지
healthier : 건강한, = healthy
prospective : 예상된, 장차
portrait : 인물사진, 흉상
provide : 제공하다, 규정하다
require : 필요하다, 요구하다
T.P.O. : Time, Place, Occasion

제4장 메이크업의 실제

1. 첫 상담

매니저 : 어서오십시오.

고객 : 안녕하세요. 오늘 처음 왔어요.

매니저 : 네, 안녕하세요. 저는 샵 매니저 그레이스입니다. 무엇을 도와드릴까요?

고객 : 메이크업을 예약하고 싶습니다.

매니저 : 네, 어떤 메이크업을 하고 싶으세요? 생각하고 오신 스타일이 있으신가요? 고객님이 원하시는 스타일에 맞춰서 전문 디자이너를 추천해 드릴 수 있습니다. 혹시 염두에 두신 디자이너가 있나요?

고객 : 아니요, 없습니다. 이곳에서는 어떤 종류의 메이크업을 받을 수 있나요?

매니저 : T. P. O 메이크업 중 뷰티 메이크업, 포토 메이크업, 웨딩 메이크업과 무대 메이크업 등이 있습니다.

고객 : 죄송하지만 잘 모르겠어요. 차이가 무엇인가요?

매니저 : 뷰티 메이크업은 일반적인 화장으로 전문가가 고객의 장단점을 보완하여 시술하는 과정입니다. 뷰티 메이크업의 용도는 가벼운 방문이나 면접, 고객과의 미팅, 모임, 맞선 등에 주로 적용되는 메이크업 스타일입니다. 포토 메이크업은 특별히 웨딩 사진 촬영이나 야외 촬영, 명함 사진, 증명사진, 이미

bridegroom. Photo makeup is exclusively for a picture taking such as wedding pictures, outdoor shots, quarter plates, portrait shots, and image shots. Wedding makeup is a wedding-related makeup offered to both the bride and the bridegroom. It includes the bride and bridegroom' s makeup for the wedding ceremony, bride' s makeup for the gift box receiving ceremony. Also makeup for the parents of the bride and bridegroom is optional. Stage makeup is a special makeup used for theaters, music concerts, musicals, artwork presentations, playing characters and musical characters.

Customer : Oh, I see. How much do you charge for each one?

Makeup artist : It differs according to your requests or circumstances. Normally, beauty makeup and photo makeup are 100,000 won respectively, and wedding makeup costs 300,000 won which includes hairstyling. The price of stage makeup differs slightly according to your needs and characters. Regular stage makeup is 200,000 won, and the price goes up if special effect makeup is required. What kind of makeup would you like to wear?

Customer : I' d like to make an appointment for Wedding Makeup. You said it includes makeup for the wedding ceremony and makeup for the gift box receiving ceremony. Can I receive just the wedding ceremony makeup?

Key words

Face shapes
① oval face
② square
③ diamond
④ triangle
⑤ inverted triangle
⑥ oblong
⑦ round

지 사진 촬영 등에 이루어지는 사진 전용 메이크업 스타일입니다. 웨딩 메이크업은 결혼에 관련된 신랑, 신부를 위한 본식 메이크업과 함맞이 메이크업, 약혼 메이크업 등으로 구분되어 신부와 신랑 모두를 위한 메이크업 과정입니다. 혼주의 메이크업은 선택 사항입니다. 무대 메이크업은 공연장이나, 음악회, 뮤지컬, 작품 발표회, 연극 캐릭터, 뮤지컬 캐릭터 등에 사용되는 특수 메이크업 과정입니다.

고객 : 네 그렇군요. 각 메이크업의 비용은 어떻게 되나요?

매니저 : 네, 고객의 요구 사항이나 경우에 따라 조금씩 차이가 있습니다. 뷰티 메이크업과 포토 메이크업은 10만원이고, 웨딩 메이크업은 헤어스타일을 포함하여 30만원입니다. 무대 메이크업은 종류와 캐릭터마다 조금씩 다르지만 일반적인 무대 메이크업은 20만원이고, 특수 분장이 추가되면 캐릭터에 따라 가격의 변동이 있습니다. 고객님은 어떤 메이크업을 원하시나요?

고객 : 저는 웨딩 메이크업을 예약하러 왔습니다. 설명 중 웨딩 메이크업에는 약혼과 함맞이를 포함하던데요, 본식 때를 위한 메이크업만 예약해도 되나요?

Makeup artist : Yes, you can.

Customer : How much does it cost?

Makeup artist : The wedding ceremony makeup is 300,000 won including hairstyling.

Customer : I see. I just want to have the wedding ceremony makeup.

Makeup artist : Okay. Are you the bride? When is your wedding?

Customer : It' s on sunday, may 16th.

Makeup artist : Where are you getting married and when does the ceremony begin?

Customer : It starts at 12 o' clock in this wedding hall.

Makeup artist : It takes about 3 hours. I think you should finish here at 11 : 40 in the morning for the wedding. Be here by 8 : 30 after having a light breakfast and wearing basic makeup.

Customer : When should the bridegroom arrive here?

Makeup artist : He should arrive here about 2 hours early after you.

Customer : Is there anything I must do before my appointment?

Makeup artist : You don' t have to prepare anything since we do everything for you concerning your makeup and hairstyling, except that you can bring one of your friends to look after your belongings. We can keep your things for you but our customers sometimes misplace their belongings in the process of preparing for their big event.

Key words

ceremony : 의식, 식, 격식
wedding ceremony : 결혼식
bride : 신부
bridegroom : 신랑

매니저 : 네, 가능합니다.

고객 : 그렇다면 본식 메이크업의 비용은 얼마인가요?

매니저 : 네, 약혼과 함맞이, 웨딩 본식은 메이크업마다 헤어를 포함해서 각각 30만원씩입니다.

고객 : 네, 그렇군요. 저는 결혼 본식을 위한 메이크업만 예약하고 싶습니다.

매니저 : 네, 신부 본인이신가요? 결혼식이 언제입니까?

고객 : 5월 16일 일요일입니다.

매니저 : 어디에서 몇 시 결혼식입니까?

고객 : 이곳 웨딩홀에서 12시입니다. 예상되는 시술 시간을 얼마 정도로 잡아야 할까요?

매니저 : 네, 시술 시간은 약 3시간 정도입니다. 신부님께서는 당일 아침 11시 40분까지 끝내야겠네요. 그렇다면 신부님께서는 아침 일찍 식사를 간단히 하시고 세안한 상태로 가벼운 기초화장만 하신 다음 저희 살롱으로 8시 30분까지 오시면 됩니다.

고객 : 신부만 메이크업 하는 시간이 그런 건가요?

매니저 : 앞서 말씀드린 대로 신랑도 같이 포함됩니다. 신랑께서는 신부님보다 2시간 정도 늦게 오시면 됩니다.

고객 : 제가 당일 따로 준비해야 할 것을 알려주세요.

매니저 : 헤어스타일과 메이크업에 관해서는 저희가 마무리 해드리니 신부님께서 따로 준비할 것은 없습니다. 단, 그날의 상황상 개인적인 소지품을 관리해 주실 친구가 한 분 오셔도 됩니다. 저희가 담당하는 소지품에 대해서는 보관해 드리지만, 큰 행사에 마음이 분주한 관계로 본인의 소지품을 분실하는 경우가 종종 있습니다. 가능하시면 고객님을 편하게 도와주실 분이 같이 오시는 게 고객님께 든든할 것 같습니다.

You' ll feel much comfortable with your friend around to help you out if possible. Do not drink too much water the day before, either. And go to bed early. It' ll make your skin healthier and make it a bit easier to apply your makeup.

Customer : Should I treat my skin before my appointment?

Makeup artist : We provide the bride a free skin care service the day before the wedding. I' d highly recommend that you come in for that.

Customer : Can I decide the time for the service? I don' t have much time.

Makeup artist : Yes, you can. But please call and let us know at least 2 days before your visit, since we could be fully booked. I' d like to remind you to have the skin care service 3 or 4 days before the wedding if possible.

Customer : Yes, I understand. You said parents makeup is optional. How much does it cost?

Makeup artist : Basically parents makeup is the same as Beauty makeup. It costs 50,000 won. Parents makeup for the wedding includes hairstyling and blow drying. The price could be at most 80,000 Won if they require an up-style.

Customer : Thank you. Is there anything I should know in addition to that?

Makeup artist : One more thing. Don' t forget to wear a button top when you come for your makeup.

Key words

decide : 결정하다
include : 포함하다
provide : 제공하다, 주다
require : 요구
ruin : 몰락, 폐허, 망치다

그 밖에는 전날 수분 섭취를 약간 자제하시고 일찍 주무시면 더 피부 건강에 좋을 듯합니다. 물론 메이크업에도 좋겠지요.

고객 : 메이크업 전에 피부 관리를 위한 준비는 어떤 것들을 해야 하나요?

매니저 : 네, 저희 살롱에서는 웨딩 메이크업에 한해서 신부님께만 무료로 전날 피부 관리를 서비스하고 있습니다. 가능하다면 꼭 오셔서 시술받도록 하십시오.

고객 : 피부 관리할 시간은 제가 정해도 될까요? 시간이 많지 않아서요.

매니저 : 네, 그렇게 하셔도 됩니다만, 혹시 예약이 되어 있을 수 있으니 오시기 이틀 전쯤에 전화주셔서 가능한 시간을 알려주시면 시간을 비워 놓겠습니다. 하지만 예약일 3~4일 안쪽으로 관리를 받으시는 게 좋습니다. 참고하십시오.

고객 : 네, 잘 알겠습니다. 그리고 혼주 메이크업은 선택 사항이라고 했는데, 어떻게 되나요?

매니저 : 혼주님 메이크업은 일단 뷰티 메이크업에 포함됩니다. 비용은 말씀드린 대로 5만원입니다. 웨딩 메이크업과 관련한 혼주 메이크업은 헤어스타일 중 드라이 시술을 포함한 가격입니다. 혼주님께서 업스타일을 원하실 경우 3만원의 추가 비용이 있습니다.

고객 : 네, 감사합니다. 더 이상 제가 알아야 할 사항은 없나요?

매니저 : 한 가지 더, 단추가 달린 상의를 입고 오시는 것을 잊지 마십시오. 이유는 메이크업과 헤어 시술이 끝나면 드레스를 입기

You wouldn' t want to ruin the finished makeup and hair when you change into your wedding gown. That' s everything you need to know.

Customer : Okay, I' ll remember that. Now I' d like to make a reservation. How much is the deposit?

Makeup artist : The deposit is 10% of the total price, and your reservation is for 7 : 30 am. on Sunday, May 16th, 2016. Can you tell me your name and number?

Customer : My name is Kristine, and my number is 02-1234-5678.

Makeup artist : Okay. Let me confirm it. Your name is Miss Kristine, and your number is 02-1234-5678. Is that correct?

Customer : Yes, that' s right. Am I all set?

Makeup artist : Yes, you are. Do you have any other questions?

Customer : Do you rent wedding gowns?

Makeup artist : Yes, we rent and customize all sorts of formal wear including wedding gowns, tuxedos and evening dresses. Do you need detailed information?

Customer : No, I was just curious. Thank you.

Makeup artist : All right. Please contact us if you would like to make any changes to your reservation or have any other questions. Here' s my card with my name and office phone number on it.

Customer : Yes, thank you. Good bye.

Makeup artist : We' ll see you for the skin care on friday May 16th. Thank you for visiting us. See you then.

Key words

correct : 맞는, 정확한
deposit : 착수금, 보증금
detail : 세부사항
formal : 격식을 차린, 정중한
tuxedo : 턱시도

위해 상의를 벗을 때 완성된 스타일을 망가뜨리지 않도록 하기 위함입니다. 더 이상 준비 사항으로 알아두실 것은 없습니다.

고객 : 잘 알겠습니다. 그럼 예약해 주십시오. 계약금을 얼마나 드려야 하나요?

매니저 : 네, 예약비용으로 전체 요금의 10%를 주시면 2016년 5월 16일 일요일 오전 5시 30분으로 예약이 됩니다. 성함과 연락 가능한 전화번호를 알려주십시오.

고객 : 제 이름은 크리스틴이고, 전화번호는 02-1234-5678입니다.

매니저 : 네, 제가 다시 한 번 확인해 드리겠습니다. 고객 성함은 크리스틴님이시고, 전화번호는 02-1234-5678입니다. 맞으신가요?

고객 : 네, 맞습니다. 그럼 예약된 건가요?

매니저 : 네, 예약되었습니다. 더 궁금하신 건 없으신가요?

고객 : 혹시, 드레스를 대여하기도 하나요?

매니저 : 네, 저희 살롱에서는 웨딩드레스와 턱시도 및 이브닝드레스 등 행사에 사용되는 각종 의상도 맞춤 및 대여해 드리고 있습니다. 필요하시면 안내해 드릴까요?

고객 : 아닙니다. 그냥 궁금해서 물어봤습니다. 감사합니다.

매니저 : 네, 그럼 예약과 관련된 변동 사항이나 그 밖의 문의 사항이 있을 시 연락 주십시오. 이것은 저희 살롱 전화번호와 제 이름이 적힌 명함입니다.

고객 : 네, 감사합니다. 안녕히 계세요.

매니저 : 그럼 5월 16일 피부 관리 시간에 다시 뵙겠습니다. 방문해 주셔서 감사합니다. 안녕히 가십시오.

2. Consultation and Treatment

Manager : Hi. You must be Miss Kristine. Nice to see you again.

Customer : I got the skin care service 2 days ago, and now I' m here for my wedding makeup.

Manager : Yes, I remember. Jenny will be designing your hair and makeup. I' ll show you the way. Please come this way.

Makeup artist : Hello, Miss Kristine.

Customer : Hi. I washed my face and hair at home.

Makeup artist : Did you wear basic makeup?

Customer : I used skin toner and lotion after washing my face as you told me last time. Is it all right?

Makeup artist : Yes, it' s fine. Please have a seat here.

Customer : My skin looks dull as I get nervous. Do you think it' ll be all right?

Makeup artist : It happens when you become sensitive with lots of things going on. But don' t worry. It' s not too serious as to affect makeup.

Customer : What do you plan to do?

Makeup artist : I' ll use nude tone makeup to produce a clean and healthy face. I' ll be using a clear and transparent base, avoiding dull heavy makeup.

Customer : What shade of eye shadow is used for nude tone makeup?

Makeup artist : A glossy pink is trendy now and it' ll be good on you for a bride. pearl powder applied to your T-zone will make you look younger and

Key words

- **accent** : 강조
- **avoiding** : 회피
- **blushing** : 얼굴을 붉힘, 얼굴이 빨개짐
- **cheerful** : 발랄한, 쾌적한
- **choice** : 선택, 선택권
- **creating** : 창조하다, 자아내다
- **dull** : 둔탁한, 칙칙한
- **heaviness** : 무거움, 중압감
- **nervous** : 불안해하는, 신경이 과민한
- **pearl powder** : 진주 가루
- **remember** : 기억하다
- **reminding** : 상기시키다
- **shade** : 그늘, 가리개
- **serious** : 심각한, 진지한
- **square** : 네모난, 사각형
- **satisfied** : 만족하는, 만족스러워하는
- **transparent** : 투명한, 명료한
- **worry** : 걱정하다, 거정하게 만들다

2. 시술 및 상담

매니저 : 안녕하세요? 지난 번 예약하셨던 크리스틴님이시죠? 다시 뵙게 되어 반갑습니다.

고객 : 이틀 전에 피부 관리를 했고 오늘은 웨딩 메이크업을 받으러 왔습니다.

매니저 : 네, 기억합니다. 제니 디자이너와 예약이 돼 있으시죠? 안내해 드리겠습니다. 이쪽으로 오십시오.

메이크업 아티스트 : 어서 오세요. 고객님!

고객 : 안녕하세요. 집에서 세안과 샴푸를 하고 왔어요.

메이크업 아티스트 : 네. 기초 화장품은 바르셨나요?

고객 : 지난 번 알려주신 대로 세안 후 스킨과 로션만 바르고 왔어요. 괜찮겠죠?

메이크업 아티스트 : 네 괜찮습니다. 메이크업 의자에 앉으실까요?

고객 : 신경을 많이 써서 그런지 피부가 칙칙한 느낌이 들어요. 괜찮을까요?

메이크업 아티스트 : 아무래도 긴장도 되고 신경 쓰실 일도 많고 해서 그럴 수 있는데 메이크업에 문제될 만큼은 아니니 걱정하지 마십시오.

고객 : 어떤 콘셉트로 메이크업을 하실 거예요?

메이크업 아티스트 : 오늘은 맑고 투명해 보이는 베이스를 선택해서 진한 화장으로 인한 칙칙함을 없애고 깨끗하고 건강한 누드톤의 메이크업을 할 겁니다.

고객 : 누드톤이라면, 아이섀도 색상은 어떤 걸 사용하게 되나요?

메이크업 아티스트 : 샤이닝 하이라이트를 표현하기 위해 요즘 신부들의 대표 색상인 글로시 핑크를 사용할 겁니다. T존 부위에 펄 파우더를 사용해 결혼식이라는

more refined. Otherwise you might look too serious and your skin might take on a darker shade.

Customer : You don' t think I may look too light in the eye of wedding guests, do you?

Makeup artist : I think you will look bright and cheerful reminding them of a rose in May, the queen of the seasons.

Customer : How about the lip color?

Makeup artist : I' m going to use pink lip gloss that goes well with the fresh eye shadow without creating heaviness. I' ll accent your young, moist and healthy lips.

Customer : I see. That' s the style I like. What about the cheek color?

Makeup artist : We' ll use pink with a little bit of a red tone to express the blushing cheek of a bride.

Customer : All right. Let' s start then.

Makeup artist : Now, what do you think of your eye shadow color? Do you like it?

Customer : Yes, I like it. It' s not done yet, right?

Makeup artist : No, you can check it again when the eyeliner and mascara are finished.

Customer : All right.

Makeup artist : Now your eye makeup' s done. What do you think?

Customer : I like it. I feel like I' m a cute bride with big eyes. Are the lips next?

자칫 엄숙하고 무거운 분위기의 신부를 발랄하고 젊음을 느낄 수 있는 세련된 신부로 표현하도록 하겠습니다.

고객 : 하객들께서 너무 가벼워 보인다고 느끼진 않겠지요?

메이크업 아티스트 : 장미를 떠올리게 하는 신부와 계절의 여왕 5월의 결혼식에 맞추어서 화사하고 밝은 분위기라고 생각하시면 됩니다. 발랄하지만 가볍지는 않은 분위기이니까요.

고객 : 입술은 어떤 색으로 하나요?

메이크업 아티스트 : 아이섀도의 화사함과 어울릴 수 있으면서 요란하지 않은 핑크톤 립글로스를 사용합니다. 젊고 촉촉하고 건강한 입술을 표현하겠습니다.

고객 : 그렇군요. 제가 좋아하는 스타일이예요. 볼 색은 어떤 걸로 하게 되나요?

메이크업 아티스트 : 약간의 붉은 톤이 들어간 핑크색 볼연지를 사용해 수줍은 신부의 볼을 연상할 수 있게 표현합니다.

고객 : 네. 좋아요. 예쁘게 해주세요.

메이크업 아티스트 : 자, 고객 지금의 아이섀도 색상은 어떠세요? 맘에 드시나요?

고객 : 네, 맘에 들어요. 다 끝난 건 아니죠?

메이크업 아티스트 : 네. 아이라이너와 속눈썹, 그리고 마스카라까지 완성되면 다시 확인하실 수 있습니다.

고객 : 네.

메이크업 아티스트 : 자 이제 아이 메이크업이 완성됐어요. 고객님 보시기에 어떠신가요?

고객 : 네 맘에 듭니다. 아주 큰 눈을 가진 귀여운 신부가 된 느낌이에요. 다음은 입술인가요?

Makeup artist : This is my choice for your lip color. What do you think?

Customer : It' s lighter than I thought. Go on.

Makeup artist : It' s going to look slightly different once you wear it. The pink looks deeper with regular makeup than you see it now. It' s because of the color of the lips themselves. So we' re going to use the lip base to make the color similar to the one you see here.

Customer : All right. I understand.

Makeup artist : Now, the lips are finished. What do you think?

Customer : I like it.

Makeup artist : Now it' s time to do the cheek and overall highlight. Is there any part you want me to take care of?

Customer : Yes. I' m a bit worried about my square chin. Can you do something about it?

Makeup artist : We can simply cover it using the cheek. But to obtain long lasting makeup for a wedding which lasts for hours, we have to consider the effects of light and shade from the basic makeup base. I have given you a foundation considering the light and shade for your base. You have nothing to worry about now. Now it' s completed. Is there any part you want to improve or don' t like?

Customer : No, it' s very nice. I' m satisfied. Thank you.

Makeup artist : Now please go to the hair-designing room to get an up-style done. The hair room is this way.

Key words

consider : 여기다, 고려하다
satisfy : 만족시키다, 충족시키다
shade : 그늘, 빛 가리개
square : 네모, 각진
worry : 걱정하다, 걱정하게 만들다

메이크업 아티스트 : 네. 입술 색상으로는 이런 색깔을 선택했습니다. 어떠세요?

고객 : 생각보다 연한 빛깔이군요. 계속해 주세요.

메이크업 아티스트 : 보기하고 바른 다음의 느낌과는 조금 차이가 있을 겁니다. 아마도 보통 화장 때는 지금 보시는 것보다 약간 진한 분홍으로 표현될 겁니다. 입술의 기본 색이 깔려 있기 때문이죠. 그래서 지금은 입술 베이스에 파운데이션 처리를 해서 보시는 색과 거의 흡사한 결과가 나올 겁니다.

고객 : 네. 그렇군요.

메이크업 아티스트 : 자 입술 화장이 완성됐습니다. 어떠십니까?

고객 : 네, 좋아요.

메이크업 아티스트 : 이젠 볼과 전체적인 하이라이트 강조가 이루어집니다. 특별히 주문하시고 싶은 부위는 없으신가요?

고객 : 네. 제가 턱이 약간 각져서 신경이 쓰여요. 보완할 수 있나요?

메이크업 아티스트 : 간단히 볼연지를 이용해 커버할 수 있지만, 결혼식이라는 장시간 행사에 유지할 수 있는 메이크업을 원하신다면, 기초화장 베이스부터 명암 처리가 이루어져야 합니다. 현재 고객의 베이스에서 충분히 명암을 고려한 파운데이션 처리를 했습니다. 걱정하지 않으셔도 됩니다. 자 이제 완성되었습니다. 더 보완을 원하시거나 맘에 들지 않는 부위가 있으시면 말씀하십시오.

고객 : 네, 아주 좋아요. 만족합니다. 고맙습니다.

메이크업 아티스트 : 다음은 업스타일을 위해 헤어디자인실로 가시면 됩니다. 감사합니다. 긴 시간 수고하셨습니다. 헤어디자인실은 이쪽입니다.

CHAPTER

05 Nail Consultation & Care

1. Hand Care

Nail artist : May I help you?

Customer : Hi. This is my first time here.

Nail artist : Well, in that case, welcome to Sharone' s Salon. I' m Kelly, the manager. What can I do for you?

Customer : I' d like to get detailed information on your nail services.

Nail artist : All right. Please have a seat. Have you ever had your nails serviced?

Customer : Yes, I had a nail tip service once. How much does it cost here?

Nail artist : Nail tip service is part of our Nail Care Program and costs 30,000 won.

Customer : I know there are several different programs. Do you have a program guide book?

Nail artist : Yes, here you are. First of all, Nail Service consists of two parts: Nail Art and Nail Care.

Customer : So the nail tip service that I had is included in Nail Care, right?

Nail artist : Yes, that' s correct. Nail care program consists of treatment for troubled nails, wet manicure and

제5장 네일 상담 및 관리

1. 손 관리

네일 아티스트 : 어서 오십시오.

고객 : 안녕하세요. 오늘 처음 왔는데요.

네일 아티스트 : 네, 샤론 살롱에 오신 것을 환영합니다. 저는 매니저 켈리입니다. 무엇을 도와드릴까요?

고객 : 네일에 대한 여러 가지 메뉴를 안내받고 싶습니다.

네일 아티스트 : 네, 이쪽으로 앉으십시오. 혹시 네일 서비스를 받아 보신 적이 있으신가요?

고객 : 네, 팁(인조손톱) 서비스를 받아 본 적이 있어요. 이곳의 요금은 얼마인가요?

네일 아티스트 : 팁 서비스는 네일 케어 부분의 작업 과정이고 요금은 3만 원입니다.

고객 : 과정에 여러 가지가 있는 걸로 알고 있는데 각 과정들에 대한 안내 책자가 있나요?

네일 아티스트 : 네, 여기 있습니다. 우선 네일 서비스의 두 가지 파트가 있는데, 한 가지는 네일 아트 부분이고 또 다른 한 가지는 네일 케어 부분입니다.

고객 : 네, 제가 받아 본 팁 서비스는 네일 아트 부분에 포함되어 있다는 거군요.

네일 아티스트 : 네, 맞습니다. 네일 케어 부분에는 손님의 이상 손톱 관리, 그 밖에도 습식 매니큐어와 페디큐어, 팁 보강

pedicure, nail tip strengthening service, nail lap, silk extension, gel nail, waxing, and acrylic. The program is used to strengthen and take care of troubled nails. Nail art program is to design your nails after the nail care using various techniques. It consists of nail enamel art, french design, stripe taping, marble art, charm art, dangle, 3D art, and air brush.

Customer : Can I get the service today?

Nail artist : Unfortunately, you need to make an appointment in advance. However let me see if anyone of our designers is available now. What service would you like to have?

Customer : I want a wet manicure.

Nail artist : Okay, you can get it today. Do you mind waiting about 10 minutes?

Customer : No, not at all.

Nail artist : Could you tell me your name and number? And I' ll make you a client card.

Customer : My name is Maria and my number is 1234-0607.

Nail artist : Let me double check your information. Your name is Ms. Maria, and your number is 1234-0607. Is that right?

Customer : Yes, that' s right.

Nail artist : Okay. Please follow me to the waiting room where you can wait comfortably. This way.

Customer : Thank you.

및 서비스, 네일 랩, 실크 익스텐션, 젤 네일, 왁싱, 아크릴릭 등이 있습니다. 문제를 가지고 있는 손톱에 보강이나 관리 개념으로 사용되는 서비스입니다. 네일 아트 부분에는 케어가 마무리된 네일에 여러 가지 테크닉을 더하여 예술적인 작품의 개념을 갖게 되는 과정입니다. 네일 에나멜 아트, 프렌치 디자인, 스트라이프 테이핑, 마블아트, 참아트, 댕글, 3D 아트, 에어브러시 등이 있습니다.

고객 : 오늘 제가 원하는 서비스를 시술받을 수 있나요?

네일 아티스트 : 죄송하게도 사전에 예약을 해야만 합니다. 그래도 혹시 오늘 시술이 가능한 디자이너를 확인해 봐 드릴까요? 오늘 받고 싶은 시술을 말씀해 주시겠어요?

고객 : 저는 습식 매니큐어를 서비스 받고 싶은데요.

네일 아티스트 : 네, 가능합니다. 10분 정도 기다리셔야 할 것 같은데 괜찮으시겠습니까?

고객 : 네, 괜찮습니다.

네일 아티스트 : 이름과 연락처를 말씀해 주시겠습니까? 고객 카드를 만들어 드리겠습니다.

고객 : 제 이름은 마리아이고, 연락처는 1234-0607입니다.

네일 아티스트 : 네, 확인해 드리겠습니다. 성함은 마리아님이고 연락처는 1234-0607입니다. 맞으신가요?

고객 : 네, 맞습니다.

네일 아티스트 : 됐습니다. 고객님, 그럼 편하게 기다리실 수 있는 대기실로 안내해 드리겠습니다. 대기실은 이쪽입니다.

고객 : 감사합니다.

2. Consultation and Treatment

Manager : Thank you for waiting. Come with me please.

Customer : Yes.

Manager : This is the designer Sherry in charge of wet manicure.

Nail artist : Hello, Ms. Alley. I’ m Sherry. Nice to meet you.

Customer : Hi. Nice to meet you, too.

Nail artist : First, sit here and relax. Please give me your left hand and tell me if you feel uncomfortable with anything.

Customer : My hands are messy after taking care of a big event. Is it all right?

Nail artist : Yes, no problem.

Customer : My hands feel very rough and dry after the hectic event. I used to have very soft hands. Do you think they’ ll get better?

Nail artist : It happens with an event because you need to take care of many things and use your hands a lot. But don’ t worry. It’ s not so serious as to affect your life.

Customer : What can I do to get my healthy skin back?

Nail artist : Today we are going to apply a hot wet compress and give you a hand massage to relieve your hand fatigue. And then a basic wet manicure will be applied to protect your nails. Your hands may not be the same as before, but it will certainly improve your condition.

Key words

AD : Anno Domini, 서기
ancient : 고대의
attractive : 매력적인
BC : Before Christ, 기원 전
comprehensive : 포괄적인
cuticle : 표피, 거스러미, (손톱뿌리를 덮은) 얇은 피부
ensure : 지키다
fatigue : 피로
free edge : 자유연(손톱 끝)
frequency : 빈도, 빈발
flexibility : 유연성
full coat : 가득 칠하다
half moon : 반월(루눌라)
messy : 지저분한, 엉망인
oblique : 완곡한, 사선의
occupation : 직업, 업무
originate : 유래하다
privilege : 특전, 특권
purchase : 구입, 구매
slim line : 가느다란
status : 신분
stiffness : 단단함, 빳빳함
suppleness : 유연성

2. 시술 및 상담

매니저 : 손님, 많이 기다리셨습니다. 이쪽으로 오십시오.

고객 : 네.

매니저 : 습식 매니큐어 담당 디자이너 셰리씨를 소개해 드리겠습니다.

네일 아티스트 : 어서 오세요. 고객님! 저는 네일 아티스트 쉐리입니다. 처음 뵙겠습니다.

고객 : 안녕하세요. 반갑습니다.

네일 아티스트 : 먼저 편안히 앉으십시오. 왼손만 제게 주시고 불편한 점이 있으시면 말씀하십시오.

고객 : 큰 행사를 치르고 와서 제 손이 지저분해요. 괜찮나요?

네일 아티스트 : 네, 괜찮습니다.

고객 : 신경을 많이 쓰고 행사를 치르느라 그런지 피부가 거칠고 건조한 느낌이 들어요. 원래 제 손은 부드럽고 좋았었는데…. 앞으로 좋아질 수 있을까요?

네일 아티스트 : 아무래도 행사를 치르다 보면 신경쓰실 일도 많고 손의 움직임도 많아서 그럴 수 있습니다. 생활에 문제될 만큼은 아니니 걱정하지 마세요.

고객 : 원래대로 좋아지려면 어떻게 하면 되죠?

네일 아티스트 : 오늘은 따뜻한 스팀 타월을 하고 손의 피로를 풀어 줄 수 있는 손 마사지를 할 겁니다. 그리고 네일을 보호할 수 있는 기초 습식 매니큐어링을 할 겁니다. 아마도 이전의 고객님 손과 똑같아지지는 않겠지만, 많이 나아지는 것을 느끼실 겁니다.

Customer : What is the first step?

Nail artist : We start with old polish removal.

Customer : Okay. What are you going to do next?

Nail artist : The rough parts of your nails will be removed to prevent future chipping and made clean. Once the basic nail care is done, the rough parts will significantly be better. How do you want me to shape your nails?

Customer : I don’ t have a specific preference, but a round shape would be good.

Nail artist : Yes, I understand. A round shape is preferred by both men and women as it’ s common and natural.

Customer : I see. I’ ve had very little nail care. I haven’ t taken care of my cuticles recently.

Nail artist : Okay, well, the amount of cuticles differs based on one’ s occupation and physical condition. The frequency of nail care you received does not play an important role.

Customer : I see. Does it mean I don’ t have to have nail care in the future?

Nail artist : No, it doesn’ t. Nail care is not just for troubled hands or nails. Hand massage relieves hand fatigue and stiffness in the hands and arms ensuring suppleness and flexibility.
Nail art can also be viewed as the finish of fashion. It helps to improve your self-esteem and gives you self-satisfaction by increasing your attractiveness to others. Therefore it’ s considered very important in modern society.

Key words

attractiveness : 매력, 애교
ensure : 반드시 ~하게 하다, 보장하다
fatigue : 피로
flexibility : 신축성 있는, 잘 구부러지는
frequency : 빈도, 빈발, 잦음
occupation : 직업, 심심풀이
relieve : 없애주다, 안도하게 하다, 완화하다
self-esteem : 자부심, 자존감
self-satisfaction : 자아만족감, 자기만족감
stiffness : 뻣뻣한, 뻑뻑한, 근육이 결리는
suppleness : 유연한, 탄력 있는
view : 견해, 관점

고객 : 어떤 시술부터 시작하는 건가요?

네일 아티스트 : 먼저 폴리시 제거부터 시작하겠습니다.

고객 : 네, 그 다음엔 어떤 시술이 들어가나요?

네일 아티스트 : 손톱의 거친 부분을 밀어내고 잘라내어 또 다른 상처를 일으키지 않고 단정하게 만듭니다. 기초 손톱관리가 들어가면 거칠어 보였던 부분이 확연히 좋아질 겁니다. 손톱의 모양은 어떻게 해드릴까요?

고객 : 특별히 선호하는 모양은 없지만, 둥근형이 좋겠어요.

네일 아티스트 : 네, 알겠습니다. 둥근형은 평범하고 자연스러워서 남녀 모두 선호하는 형태입니다.

고객 : 그렇군요. 손톱관리를 거의 하지 않았네요. 큐티클(거스러미) 정리도 거의 하지 못했어요.

네일 아티스트 : 네. 손님마다 차이가 있지만, 손질의 유무를 떠나 거스러미의 양도 직업과 체질에 따라 다르기 때문에 손질의 탓만 할 수는 없습니다.

고객 : 그렇군요. 그럼 저는 앞으로 네일 케어를 받지 않아도 되나요?

네일 아티스트 : 그렇진 않습니다. 굳이 손의 문제가 생기거나 손톱의 문제가 발생해야만 네일 케어를 받는 것은 아닙니다. 네일 케어를 통한 손 마사지는 손의 피로회복과 손과 팔의 근육을 풀어 주어 유연성과 탄력성을 가지게 합니다.

네일 아트를 패션의 마무리 개념으로 생각할 수도 있으며, 타인에게 매력적으로 보여 호감을 주고 자신에게 더 만족감을 주어 자아 존중감을 높여주는 역할도 합니다. 그래서 네일아트는 현대 사회에서 아주 중요하게 여겨지고 있지요.

Customer : Oh, I see. When did nail art start?

Nail artist : It has a very long history. First of all, the term "manicure" originated from Latin words "manus (hand)" and "cura (care)", which were then combined. "Manicure" refers to a comprehensive nail care including nail shaping, cuticle removal, hand massage, and coloring. It has developed from 3,000 B.C. to today continuing for 5,000 years.

Customer : Oh, I see. It has developed for a very long period of time.

Nail artist : Yes. According to literature, nail care began in Egypt and China around 3,000 B.C. catering specifically to the upper class. Colors were extracted from bushes or plants at that time. In ancient Egypt, nail colors were considered as one of the important symbols of social status : Royalty used dark colors while low class people were allowed to use only light colors.

Customer : All right. It' s very interesting to know the fact that only royalty and the nobility enjoyed nail care.

Nail artist : The more interesting thing is that nail care spread among men in medieval times when so many wars occurred. Medieval soldiers, unlike ancient people, applied the same color on their lips and nails using dyes before going to a war.

Customer : Oh, I see. Nail care has a very interesting history.

Key words

- **according to** : ~에 따르면, ~에 의하면
- **allow** : 허락하다, 용납하다
- **ancient** : 고대의
- **bush** : 관목, 덤불
- **low class** : 하류층
- **nobility** : 고귀함, 고결함
- **medieval** : 중세의
- **period** : 시대, 시기
- **refer to** : ~와 관련이 있다
- **royalty** : 귀족의
- **social statue** : 사회적 지위, 신분
- **spread** : 퍼지다, 펼치다
- **symbols** : 상징, 표시
- **upper class** : 상류층

고객 : 아, 그렇군요. 네일 아트는 언제부터 시작된 건가요?

네일 아티스트 : 네일 아트는 아주 오랜 역사를 가지고 있습니다. 우선 우리가 흔히 상식처럼 알고 있는 '매니큐어'란 용어의 원리부터 알아보면, 라틴어의 '마누스(manus/손)'와 '큐라(cura/관리)'에서 매니큐어라는 단어가 파생되어 손톱의 모양 정리, 큐티클의 정리, 손 마사지, 컬러링 등을 포함한 총괄적인 손의 관리를 뜻합니다. 기원전 3,000년부터 지금까지 발전되어 왔습니다. 그러니 5,000년에 걸쳐 네일 아트가 이어진 것입니다.

고객 : 아, 그렇군요. 오랜 시간 동안 발전해 왔네요.

네일 아티스트 : 네, 문헌에 따르면 기원전 3,000년 경 이집트와 중국의 상류층에서 최초로 네일 관리가 시작된 것으로 전해집니다. 기원전에는 관목이나, 식물 등에서 색상을 추출했어요. 고대 이집트에서는 네일의 색깔이 사회적인 지위를 나타내는 중요한 기준의 하나로 여겨져 왕족은 짙은 색으로, 낮은 계층의 사람들은 옅은 색만을 사용하게 하여 신분을 구분하였습니다.

고객 : 그랬군요. 왕족이나 귀족에게만 네일 관리가 이루어졌다니 무척 흥미로운 일이군요.

네일 아티스트 : 더 흥미로운 사실은 수많은 전쟁이 치러진 중세시대에 네일관리가 남자들 사이에 퍼졌다는 겁니다. 고대와 달리 전쟁터에 나가는 중세 군인들은 전쟁에 나가기 전에 염료를 사용해서 입술과 네일에 같은 색을 칠하기도 했다고 합니다.

고객 : 아, 그래요. 네일관리는 무척 흥미로운 역사를 가지고 있군요.

Nail artist : Yes, but at the beginning of early modern days, this privilege started to spread among common people promoting the development of various nail care products. Today nail techniques and nail products are specialized and subdivided compared to early modern times, and the nail industry is developing into one of the main businesses worldwide. Isn' t the historical changes of nail care very fascinating?

Customer : It' s very interesting and refreshing to learn about nail care.

Nail artist : Now the hand massage has finished. Please choose an enamel color you want to use.

Customer : How about pink for spring? Do you think it suits me?

Nail artist : Yes, it will go well with your fair skin. Do you have a particular line you prefer? Half moon, slim line, free edge, or full coat?

Customer : I prefer the slim line because my nails are wide.

Nail artist : All right. Then I will design a simple nail art by applying bright colored oblique stripes on the slim lined pink enamel.

Customer : Oh! It' s very pretty! Enamel comes off easily as I do housework. How can I maintain this pretty nail art for a long time?

Nail artist : Just apply this Top Coat I' m using now on the enamel nails from time to time, and the enamel will last for a long time. If I may add, adjust the frequency of application according to your use of hands.

Key words

come off : 떼어낼 수 있다, 제거할 수 있다
copare to : ~와 비교하다
fascinating : 대단히 흥미로운, 매력적인
maintain : 유지하다
privilege : 특전, 특권
subdivide : 다시 나누다, 세분하다

네일 아티스트 : 네. 그러나 근대에 접어들어 상류층의 특권으로만 여겨졌던 종전과는 달리 일반인에게 대중화되기 시작되면서 다양한 네일 제품이 개발되었어요. 현대에는 네일 테크닉과 네일 제품이 근대보다 전문화 · 세분화되었으며 네일 산업이 세계의 주요 산업의 하나로 발돋움하고 있답니다. 네일의 역사적 변천 과정이 무척 흥미롭죠?

고객 : 네일 관리가 무척 흥미롭고 새롭게 느껴지네요.

네일 아티스트 : 이제 손 마사지가 끝났습니다. 좋아하시는 색깔의 에나멜을 선택해 주세요.

고객 : 봄맞이용 핑크는 어떨까요? 제게 잘 어울릴까요?

네일 아티스트 : 네, 피부톤이 밝은 편이어서 잘 어울리실 거예요. 특별히 선호하는 라인이 있으신가요? 하프문(반월부분), 슬림라인(가늘게), 프리엣지(자유연/손톱 끝 부분), 풀코트(손톱전체 가득 칠하기) 등이 있어요.

고객 : 전 제 손톱이 넓어서 슬림라인의 컬러링을 원해요.

네일 아티스트 : 네. 그렇다면 핑크색 에나멜 슬림라인 위에 기본적인 밝은 색조의 스트라이프 사선 기법으로 간단한 네일아트를 시술해 드리겠습니다.

고객 : 오! 아주 멋진데요. 하지만 집에서 일하다 보면 에나멜이 쉽게 지워져요. 예쁜 네일 아트를 오래 유지하려면 어떻게 해야하나요?

네일 아티스트 : 지금 제가 바르는 탑코트를 에나멜이 칠해진 네일 위에 가끔 발라 주시면 디자인한 에나멜을 오랫동안 유지할 수 있습니다. 참고로 덧바르는 횟수는 손의 쓰임이 많고 적음에 따라 조절해 주시면 됩니다.

Customer : I see. Can I buy Top Coat here?

Nail artist : Of course. We are stocked with basic nail care tools. You can purchase self-nail care items at the counter. Now it' s all done!

Customer : My hands look very tidy and pretty, and I feel very refreshed as well.

Nail artist : If you get your nails treated regularly, it will take a much shorter time the next time you visit. And it will be more pleasant with extra massage time.

Customer : I see. Can I come 2 weeks later? I' m very busy next week.

Nail artist : Sure, let me check our schedule. Yes, you can. You can make anappointment for 4pm. on Wednesday, July 2nd. Is it okay?

Customer : Yes, that' s fine.

Nail artist : Your reservation is done.

Customer : Thank you. Bye.

Nail artist : Thank you. See you on Wednesday two weeks later.

고객 : 그렇군요. 여기서 탑코트를 살 수 있나요?

네일 아티스트 : 물론입니다. 네일 케어를 위한 기본 도구를 갖추고 있어요. 홈케어에 필요한 물품은 카운터에서 구입하실 수 있습니다. 자! 완성되었습니다.

고객 : 아주 깔끔하고 보기도 좋은데다 기분도 한결 상쾌해진 느낌이에요.

네일 아티스트 : 정기적으로 네일 케어를 받으시면 다음 방문때는 지금 소요된 시간보다 훨씬 단축될 수 있습니다. 게다가 마사지 시간이 추가되어 기분 좋은 케어를 받으실 수 있을 겁니다.

고객 : 그렇군요. 혹시 2주일 후에도 예약이 가능한가요? 다음 주는 제가 무척 바쁘거든요.

네일 아티스트 : 확인해 드리겠습니다. 네, 가능합니다. 7월 2일 첫째 수요일 4시로 예약이 가능한데 괜찮으시겠습니까?

고객 : 네, 좋습니다.

네일 아티스트 : 예약되었습니다.

고객 : 감사합니다. 안녕히 계세요.

네일 아티스트 : 네, 감사합니다. 2주 뒤 수요일에 뵙겠습니다.

3. Manicure

Key words

advanced : 고급의
attention : 주의
attire : 의복, 복장
chip : 조작, 부스러기
complexion : 안색, 양상
daily : 매일, 나날의
difficulty : 어려움, 곤경, 장애
draw : 그리다. 끌어당기다
favorite : 매우 좋아하는, 마음에 드는
gorgeous : 아주 멋진, 선명한, 화려한
leather : 가죽
length : 길이
penetrate : 뚫고 들어가다, 관통하다
protect : 보호하다
really : 정말로, 진심으로
roll : 감다
seal : 봉하다, 밀봉하다
selection : 선택, 선발, 선정
smear : 마구 바르다
sophisticated : 교양 있는, 수준 있는, 지적인
splash : 떨어지다
subtle : 미묘한, 절묘한
translucent : 반투명한
almond shape : 아몬드 형
oval shape : 타원형, 계란형
rectangular (square) shape : 직사각형, 네모형
round shape : 둥근형
round square : 둥근 사각형

Nail artist : Mrs. Smith, we have two kinds of nail care programs, a basic and a special manicure program. The special consists of two more services : removal of dead skin cells on your hands, and a wax or a moisture pack depending on the condition of your hands. Your hands are very warm and dry. We' d strongly recommend you to have the special care program including the moisture pack.

Customer : Ok. That sounds good.

Nail artist : You might be a little uncomfortable while your hands are being treated. Do you need to get anything before we get started? The nail remover might splash and get on your leather things, if possible, don' t keep anything on the table.

Customer : I see. Thank you.

Nail artist : First, I' ll remove your colored nail polish.

Customer : Okay.

Nail artist : How long do you want your nails? Would you like them rounded or squared?

Customer : I' m satisfied with this length. I just want to change it into a rectangular shape.

Nail artist : Sure. If you are uncomfortable, just let me know. I' ll care for your cuticles now. You might feel a little pain, please let me know if you feel any pain.

Customer : Okay.

3. 매니큐어

네일 아티스트 : 고객님. 저희 샵에는 기본 매니큐어와 스페셜 매니큐어프로그램이 있습니다.스페셜 매니큐어는 기본 매니큐어에 각질제거와 고객님의 손 상태에 따라 파라핀 또는 수분팩이 추가되는 프로그램입니다. 고객님은 건조하고 열이 많은 손입니다. 고객님께 수분팩이 포함된 스페셜 매니큐어를 추천해 드리고 싶습니다.

고객 : 네, 그럼 추천해주신 스페셜 매니큐어를 하겠습니다.

네일 아티스트 : 관리가 시작되면 손사용이 불편하오니, 필요하신 물건이 있으시면 미리 꺼내 놓으세요. 혹시, 관리 도중 리무버가 튈 수 있으니 가죽소재 물건을 가급적 테이블위에 올려놓지 마시고요.

고객 : 아, 그럴 수도 있겠네요. 고마워요

네일 아티스트 : 컬러부터 지워드리겠습니다.

고객 : 네.

네일 아티스트 : 손톱길이는 어느 정도로 해드릴까요? 손톱모양은 둥글게 해드릴까요? 아니면, 사각형 모양으로 해드릴까요?

고객 : 길이는 지금이 적당해요. 모양만 사각형으로 바꿔 주세요.

네일 아티스트 : 알겠습니다. 시술 중 불편하신 점 있으시면 말씀해 주세요. 큐티클 정리하겠습니다. 큐티클은 예민한 부분이라 관리 도중 통증이 느껴질 수 있으니 통증이 느껴지면 참지마시고 말씀해 주세요.

고객 : 네.

Nail artist : Okay, I' m finished. Before applying the moisture pack, I' ll remove the dead cells on your hands. May I roll up your sleeves?

Customer : Of course.

Nail artist : The removal is finished. Now I' ll wipe your hands with a warm towel before I give you a massage. the towel is just warm, don' t worry.

Customer : Thank you.

Nail artist : I' ll give you a hand massage now. Please let me know if the massage is too strong or weak.

Customer : Okay, I' ll do.

Nail artist : You' re all set. You are supposed to have a twenty-minute moisture pack. You can have a difficulty using the bathroom while the moisture pack and coloring is being applied. Do you want to use the bathroom now?

Customer : Okay, sure, I' ll use the bathroom now.

Nail artist : I' m going to put the moisture pack on your hands. It' s a little cold.

Customer : Okay, no problem.

Nail artist : I' ll remove the pack now. How was it? Can you feel any difference?

Customer : Yes, my hands are really moist. I' d like them to be like this always.

Nail artist : Do your best to apply some hand cream on your hands daily.

Customer : Okay, I' ll do that.

Nail artist : Now we can begin to color them. Do you want any special color or do you have a favorite color?

Key words

attire : 의복, 복장
complexion : 안색, 양상
favorite : 마음에 드는, 매우 좋아하는
seal : 봉하다, 밀봉하다
smear : 마구 바르다
translucent : 반투명한

네일 아티스트 : 큐티클 정리가 다 끝났습니다. 수분팩을 하기 전에 각질제거를 해드리겠습니다. 소매를 걷어도 될까요?

고객 : 네. 그러세요.

네일 아티스트 : 각질제거가 끝났습니다. 마사지 해드리겠습니다. 마사지 전 온타월로 손을 닦아드리겠습니다. 온타올 온도 체크를 해보니 무리하게 뜨겁지 않더라고요. 안심하세요.

고객 : 고마워요.

네일 아티스트 : 마사지 해드리겠습니다. 마사지가 약하거나 강하여 불편하시면 말씀해 주세요.

고객 : 네. 알겠습니다.

네일 아티스트 : 마사지가 끝났습니다. 20분 정도 수분팩을 해드릴 예정입니다. 수분팩부터 컬러링까지 시술하는 동안은 손사용이 불편하오니 지금 화장실에 미리 다녀오시겠어요?

고객 : 네. 그럴게요. 그게 좋을 듯하네요.

네일 아티스트 : 수분팩을 얹어 드리겠습니다. 수분팩은 차가우니 놀라지 마세요.

고객 : 네.

네일 아티스트 : 팩을 제거해드리겠습니다. 어떠세요? 변화가 느껴지세요?

고객 : 네. 손이 촉촉해졌네요. 제 손이 늘 이랬으면 좋겠어요.

네일 아티스트 : 핸드크림을 수시로 발라주시면 좋으실겁니다.

고객 : 네. 노력해 볼게요.

네일 아티스트 : 컬러링에 들어가겠습니다. 특별히 원하시거나 평소에 즐기시는 컬러가 있으신가요?

Customer : No, not really. But I do have an important meeting this week and would like a sophisticated but subtle color. I don' t want to draw much attention to myself.

Nail artist : I see. Let me show you our color selection.

Customer : Great.

Nail artist : This is a brand new color that has recently been in fashion. It' s a pinkish translucent that goes well with any complexion and attire.

Customer : It' s absolutely gorgeous! It' s just my shade. May I test it?

Nail artist : Of course! What do you think?

Customer : It' s great. Thanks for recommending it.

Nail artist : First, I' ll apply a basic coat to protect your nails. Now, I' ll apply the color we choose.

Customer : OK.

Nail artist : Now, I' ll apply the top coat to seal your nails.

Customer : Yes.

Nail artist : OK, you' re all set. Please sit here to dry your nails. Be careful when you move, or you could smear or chip them.

Customer : Sure.

Nail artist : I' ll let your nails dry with a cool wind under the dryer for about 15 minutes. If you feel anything uncomfortable, let me know.

Customer : I see.

고객 : 아니요. 그런 건 없어요. 그런데, 제가 이번 주에 회사에서 중요한 행사가 있어요. 화려하지 않으면서 세련된 색을 원해요. 너무 이목을 끌지 않으면 좋겠네요.

네일 아티스트 : 아~ 그러세요. 알겠습니다. 그러면 제가 추천해보겠습니다.

고객 : 네, 좋아요.

네일 아티스트 : 이 컬러는 이번 주에 새로 들어온 신상품입니다. 핑크빛이 살짝 감도는 반투명 복숭아색이라서 피부색, 의상에 관계없이 무난하게 코디하실 수 있으실 거예요.

고객 : 어머~ 예쁘네요. 한번 테스트해볼 수 있을까요?

네일 아티스트 : 네. 제가 도와드리겠습니다. (테스트 후) 어떠세요?

고객 : 잘 어울리네요. 이걸로 해주세요.

네일 아티스트 : 네, 알겠습니다. 손톱을 보호하기 위해 베이스코트부터 발라드리겠습니다.

고객 : 네.

네일 아티스트 : 컬러 발라드리겠습니다. 컬러 코팅제인 탑코트 발라드리겠습니다.

고객 : 네.

네일 아티스트 : 모든 시술이 끝났습니다. 컬러가 잘 마를 수 있도록 드라이기로 자리를 옮겨 드리겠습니다. 컬러가 찍힐 수 있으니 이동시 조심하시고요.

고객 : 알겠어요.

네일 아티스트 : 컬러가 잘 마르도록 찬바람으로 해드리겠습니다. 15분 정도 말려 주세요. 말리는 중에 불편하신 점이 있으시면 말씀해 주세요.

고객 : 네.

Nail artist : All done. I' ll apply some oil to your cuticles. Please don' t wipe them until the oil penetrates smoothly.

Customer : Sure. Can I pay now? How much do I owe you?

Nail artist : Ok. It comes to 35,000 won. May I help you pay so you don' t chip the color on your nails ?

Customer : Thank you. Next time I' ll ay in advanced.

Nail artist : I should have said that. I' m sorry.

Customer : That' s all right. Thanks for your help today.
Nail artist : My pleasure. See you soon.

네일 아티스트 : 다 끝나셨습니다. 큐티클부분에 오일 좀 발라 드리겠습니다. 오일이 자연스럽게 흡수되게 닦아내지 마세요.

고객 : 네. 알겠습니다. 계산 도와주세요. 오늘 관리 받은 게 얼마죠?

네일 아티스트 : 네. 알겠습니다. 오늘 관리 금액은 35,000원입니다. 컬러가 손톱에 묻을 수 있으니 제가 계산을 도와드려도 될까요?

고객 : 네. 부탁드리겠습니다. 다음에는 관리 전에 돈을 미리 꺼내놔야겠네요.

네일 아티스트 : 제가 관리 전 미리 안내해 드렸어야 했는데 죄송합니다.

고객 : 괜찮아요. 오늘 수고 많으셨습니다.

네일 아티스트 : 저희 샵을 이용해 주셔서 감사합니다. 다음에 또 뵙겠습니다.

4. Pedicure

Nail artist : We have a basic and a deluxe pedicure treatment. The deluxe pedicure service includes the removal of dead skin from the hard sole and a calf massage along with basic treatment. I' ll check your feet now. Your skin is dry and cracked. You have some calluses and dead skin. I would recommend that you have the deluxe service including the dead skin cell removal and calf massage.

Customer : Today my feet feel heavy and tired. I am suffering from foot fatigue. I' ll try the deluxe service.

Nail artist : Before the treatment, I would recommend using the bathroom, as you won' t be able to go during the service. Also if you have anything in your pockets or bag that you might need during the service, please take them out now.

Customer : Ok, thank you. I' ll go to the bathroom now.

Nail artist : Okay, take your time.

Nail artist : (after using the bathroom) You can roll up your pants or change into a gown. Which do you prefer?

Customer : I' ll roll up my pants. That' s Okay.

Nail artist : I' ll remove the old pedicure.

Customer : yeah.

Nail artist : What length do you want your toe nails? I' ll trim and square your toenails to prevent them from digging into your skin.

Key words

- **absorb** : 흡수하다
- **along** : ...을 따라
- **calf** : 종아리, 송아지
- **calluses** : 손발의 굳은 살
- **check** : 살피다
- **cracked** : 금이 간, 갈라진
- **cracking** : 가라지는 소리
- **deluxe** : 호화스러운, 사치스러운, 고급의
- **dig** : 파다
- **discomfort** : 불편, 불편하게 하는 것
- **hesitate** : 망설이다, 거리끼다
- **hope** : 희망하다, 바라다
- **lukewarm** : 미지근한
- **match** : 성냥, 경기, ...에 맞추다
- **pretty** : 아주, 매우, 매력적인, 예쁜
- **rejuvenate** : 젊어 보이게 하다, 활기를 되찾다
- **rip** : 찢다, 떼어내다
- **slight** : 약간의, 경미한
- **suffering** : 고통, 괴로움
- **swimsuit** : 수영복
- **toenails** : 발톱

4. 패디큐어

네일 아티스트 : 고객님, 저희 샵에는 기본 패디큐어와 스페셜 패디큐어프로그램이 있습니다. 스페셜 패디큐어는 기본 패디큐어에 발바닥 각질제거와 종아리 마사지가 추가되는 프로그램입니다. 고객님의 발을 체크해 보겠습니다. 고객님은 발바닥이 건조하시고 굳은 살과 각질이 있는 편입니다. 고객님께 각질제거와 종아리 마사지가 포함된 스페셜 패디큐어를 추천해 드리고 싶습니다.

고객 : 네, 오늘은 발이 너무 피로하네요. 추천해주신 스페셜 패디큐어를 해보겠습니다.

네일 아티스트 : 관리를 시작하시기 전에 관리 중 이동이 불편하오니, 화장실에 미리 다녀오셔도 좋을 것 같습니다. 혹시, 필요하신 물건이 있으시면 미리 꺼내 놓으세요.

고객 : 아... 그럴 수도 있겠네요. 고마워요. 미리 화장실에 다녀오겠습니다.

네일 아티스트 : 네, 다녀오세요.

네일 아티스트 : (화장실 다녀온 후) 바지는 무릎 위까지 걷겠습니다. 혹시 바지에 구김이 갈 수 있으니, 저희가 준비해 놓은 관리복으로 바꾸어 입으셔도 됩니다.

고객 : 아니요, 괜찮습니다.

네일 아티스트 : 컬러부터 지워드리겠습니다.

고객 : 네.

네일 아티스트 : 발톱길이는 어느 정도로 해드릴까요? 발톱모양은 파고드는 발톱을 방지하기 위해서 사각형 모양으로 해드리겠습니다.

Customer : My stockings rip easily. I don’ t know the exact reason, but it can be that my nails are too long, therefore, please trim them a bit.

Nail artist : I see. Let me know if you feel any discomfort.

Customer : Sure.

Nail artist : I’ ll care for your cuticle now. You might feel a slight pain. Even if it is really slight, don’ t hesitate to tell me. The cuticle is a very sensitive area.

Customer : I see.

Nail artist : I’ ve done caring for your cuticles. I’ ll remove the hard skin from your sole.

Customer : Thank you.

Nail artist : OK, it’ s done. Now I’ ll let your feet soak for a few minutes. I put some aroma oil into the water to help rejuvenate your foot. It’ ll make your feet feel fresh. I’ ll check the temperature of the water first. If it’ s too hot or cold, let me know.

Customer : Sure. It is a little hot for me.

Nail artist : Okay, sorry about that. I’ ll adjust the temperature again. I’ ll check it out once more. How is it now?

Customer : Great.

Nail artist : I’ ll start to soak your feet in hot water. You need about 20 minutes. for your feet to rest.

Customer : Okay.

Nail artist : Your foot bath is done. after I rinse your feet, I ‘ll massage them.

Customer : Yes.

Key words

rejuvenate : 다시 젊어 보이게 하다, 활기를 되찾게 하다

soak : 담그다, 흠뻑 적시다

고객 : 길이가 길어서인지 스타킹에 자꾸 구멍이 나네요. 짧게 해주세요.

네일 아티스트 : 네. 알겠습니다. 시술 중 불편하신 점이 있으면 말씀해주세요.

고객 : 네.

네일 아티스트 : 큐티클 정리하겠습니다. 큐티클은 예민한 부분이라 관리 도중 통증이 느껴질 수 있으니 통증이 느껴지면 참지마시고 말씀해주세요.

고객 : 네.

네일 아티스트 : 큐티클 정리가 다 끝났습니다. 발바닥 각질제거를 해드리겠습니다.

고객 : 네. 그러세요.

네일 아티스트 : 각질제거가 끝났습니다. 족욕을 해드리겠습니다. 발이 피로하시다고 하셔서 피로회복에 도움이 될 아로마 오일을 넣었습니다. 피로회복에 도움이 되셨으면 좋겠네요. 물 온도 테스트 해드리겠습니다. 불편하시면 말씀해주세요.

고객 : 네. 알겠습니다. (테스트 후) 조금 뜨거운 것 같아요.

네일 아티스트 : 죄송합니다. 온도를 다시 맞춰드리겠습니다. 다시 한 번 온도 테스트 해드리겠습니다. 어떠세요?

고객 : 네. 적당한 것 같아요.

네일 아티스트 : 족욕을 시작하겠습니다. 발의 피로를 풀기 위해 족욕은 20분 정도 할 예정입니다.

고객 : 네.

네일 아티스트 : 족욕이 끝났습니다. 발을 헹군 후 마사지 해드리겠습니다.

고객 : 네.

Nail artist : I' ll dry your feet first before the massage. The towel is just lukewarm, don' t worry.

Customer : OK. Thank you.

Nail artist : I' ll start the foot massage. If you think it is too strong or too weak for you, let me know, please.

Customer : Sure.

Nail artist : The foot massage is done. Before you go to sleep tonight, try to apply some foot cream. This will help keep your feet smooth and moist. It will also prevent cracking.

Customer : Thank you. I' ll do that.

Nail artist : I' ll start coloring them now. Do you have any special or favorite color you like?

Customer : Not really, but I am going to the beach this week. I' d like to match the color with my white swimsuit. But I don' t want all my toenails to be white.

Nail artist : I see. How about a French polish, the same color as that of your swimsuit? Your toenails have a beautiful shape and look healthy, so the French polish will suit you well.

Customer : May I test it?

Nail artist : Of course I' ll help you. (after the test) Do you like it?

Customer : It' s really pretty. I' ll take it.

Nail artist : Great. I' ll apply the base coat first in order to protect your toenails.

네일 아티스트 : 마사지 전 온타월로 발을 닦아드리겠습니다. 온타올 온도 체크를 해보니 무리하게 뜨겁지 않더라고요. 안심하세요.

고객 : 네. 고맙습니다.

네일 아티스트 : 마사지 해드리겠습니다. 마사지가 약하거나 강하여 불편하시면 말씀해 주세요.

고객 : 네. 알겠습니다.

네일 아티스트 : 마사지가 끝났습니다. 오늘 관리 후, 댁에 돌아가셔서 주무시기 전에 풋크림을 다시 한 번 발라주세요. 부드러운 발을 유지하시는 데 많은 도움이 될 것입니다.

고객 : 네. 감사해요. 그렇게 할게요.

네일 아티스트 : 컬러링에 들어가겠습니다. 특별히 원하시거나 평소에 즐기시는 컬러가 있으신가요?

고객 : 아니요. 그런 건 없어요. 그런데, 제가 이번 주에 바다로 여행을 갈 예정입니다. 제 수영복이 흰색 비키니인데, 수영복 색에 맞추고 싶은데... 발톱 전체를 흰색으로 바르는 건 조금 망설여지고요.

네일 아티스트 : 아~ 그러세요. 알겠습니다. 그러면 제가 추천해보겠습니다. 고객님의 수영복 색상과 같은 흰색으로 프렌치를 해보시는 건 어떠세요? 고객님의 발톱은 워낙 모양도 예쁘고 발톱의 혈색도 건강해서 프렌치를 하시면 잘 어울리실 것 같아요.

고객 : 한번 테스트해 볼 수 있을까요?

네일 아티스트 : 네. 제가 도와드리겠습니다. (테스트 후) 어떠세요?

고객 : 예쁘네요. 이걸로 해주세요.

네일 아티스트 : 네 알겠습니다. 발톱을 보호하기 위해 베이스코트부터 발라드리겠습니다.

Customer : Yes.

Nail artist : I' ll paint the color.

Customer : Yes.

Nail artist : I' ll apply the top coat and seal them now.

Customer : Yes.

Nail artist : All services are done. Please have a seat here to dry your nails. The color is still setting and can be chipped. Be careful when you move.

Customer : Yes I see.

Nail artist : I' ll let dry your nails for their perfect dry. Fifteen minutes. will be enough. Feel free to let me know if you feel any discomfort.

Customer : Yes.

Nail artist : All done now. I' ll apply some oil to the cuticle area. Please be careful so you don' t wipe off the polish. It needs to absorb naturally. Also take care when walking, because of the massage cream on your sole.

Customer : I see. I' ll pay now. Can I have the bill? How much is everything?

Nail artist : (at the register) You had a pedicure, a manicure, a foot massage. Is that right? The total comes to fifty-five thousand won.

Customer : Alright.

Nail artist : Thanks for coming in today. I hope to see you soon.

고객 : 네.

네일 아티스트 : 컬러 발라드리겠습니다.

고객 : 네.

네일 아티스트 : 컬러 코팅제인 탑코트 발라드리겠습니다.

고객 : 네.

네일 아티스트 : 모든 시술이 끝났습니다. 컬러가 잘 마를 수 있도록 드라이기로 자리를 옮겨드리겠습니다. 컬러가 찍힐 수 있으니 이동시 조심하시고요.

고객 : 네.

네일 아티스트 : 컬러가 잘 마르도록 찬바람으로 해드리겠습니다. 15분 정도 건조해주세요. 건조 중에 불편하신 점이 있으시면 말씀해주세요.

고객 : 네.

네일 아티스트 : 다 끝났습니다. 큐티클 부분에 오일 좀 발라드리겠습니다. 오일이 자연스럽게 흡수되게 닦아내지 마세요. 혹시 마사지 크림으로 인해 발바닥이 미끄러우실 있으니 조심하시고요.

고객 : 네. 알겠습니다. 계산 도와주세요. 오늘 관리받은 비용이 모두 얼마죠?

네일 아티스트 : 네. 알겠습니다. 패디큐어와 매니큐어, 발 마사지까지 맞으시죠? 오늘 관리 금액은 55,000원입니다.

고객 : 네. 오늘 수고 많으셨습니다.

네일 아티스트 : 저희 샵을 이용해 주셔서 감사합니다. 다음에 또 뵙겠습니다.

Appendix

부 록

Business English for Beauticians

CHAPTER 01

색표(Color Chart)

관용색명	먼셀표기	영문색명	관용색명	먼셀표기	영문색명
벚꽃색	2.5R 9/2	cherry blossom	호박색(채소)	10YR 7/14	pumpkin
상록수색	10G 3/8	evergreen	카네이션핑크	2.5R 8/6	carnation pink
황토색	10YR 6/10	yellow ocher	피콕그린	7.5BG 3/8	peacock green
루비색	2.5R 3/10	ruby	황갈	10YR 5/10	yellowish brown
청록	10BG 3/8	blue green	크림슨	2.5R 3/10	crimson
호두색	10YR 5/6	walnut	물색	5B 7/6	aqua blue
베이비핑크	5R 8/4	baby pink	점토색	10YR 4/4	clay
하늘색	7.5B 7/8	sky blue	연지색	5R 5/12	madder red
세피아	110YR 2/2	sepia	시안	7.5B 6/10	cyan
딸기색	5R 4/14	strawberry	베이지	2.5Y 8.5/4	beige
세룰리안블루	7.5B 4/10	cerulean blue	카민	5R 4/12	carmine
해바라기색	2.5Y 8/14	sunflower	파스텔블루	10B 8/6	pastel blue
장미색	5R 3/10	rose	노른자색	2.5Y 8/12	yolk yellow
파우더블루	10B 8/4	powder blue	자두색	5R 3/10	plum
금발색	2.5Y 7/6	blond	스카이그레이	10B 8/2	sky gray
와인레드	5R 2/8	wine red	모래색	2.5Y 7/2	sand
박하색	2.5PB 9/2	mint	복숭아색	7.5R 8/6	peach
베이지그레이	2.5Y 7/1	beige gray	파랑	2.5PB 4/10	blue
산호색	7.5R 7/8	coral	카키색	2.5Y 5/4	khaki
프러시안블루	2.5PB 2/6	prussian blue	빨강	7.5R 4/14	red
청동색	2.5Y 4/4	bronze	인디고블루	2.5PB 2/4	indigo blue
토마토색	7.5R 4/12	tomato	모카색	2.5Y 3/4	mocha
비둘기색	5PB 6/2	dove	사과색	7.5R 3/12	apple

크림색	5Y 9/4	cream	코발트블루	5PB 3/10	cobalt blue
석류색	7.5R 3/10	pomegranate	연미색	5Y 9/2	mayonnaise
사파이어색	5PB 3/6	sapphire	새먼핑크	10R 7/8	salmon pink
상아색	5Y 9/2	ivory	감(紺)색	5PB 2/4	navy blue
주색	10R 5/16	vermilion	우유색	5Y 9/1	milk
라벤더색	7.5PB 7/6	lavender	적갈	10R 3/10	reddish brown
노랑	5Y 8.5/14	yellow	군청	7.5PB 2/8	ultramarine blue
주황	2.5YR 6/14	orange	개나리색	5Y 8.5/14	forsythia
남색	7.5PB 2/6	bluish violet	당근색	2.5YR 6/12	carrot
바나나색	5Y 8/12	banana	라일락색	5P 8/4	lilac
감색(과일)	2.5YR 5/14	persimmon	겨자색	5Y 7/10	mustard
보라	5P 3/10	purple	구리색	2.5YR 6/14	copper
레몬색	7.5Y 8.5/12	lemon	포도색	5P 3/6	grape
코코아색	2.5YR 3/4	cocoa	참다래색((키위색)	7.5Y 4/6	kiwi
마젠타	5RP 5/14	magenta	살구색	5YR 8/8	apricot
황녹색	10Y 6/10	yellowish green	진달래색	7.5RP 5/12	azalea
갈색	5YR 4/8	brown	올리브색	10Y 4/6	olive
자주	7.5RP 3/10	reddish purple	밤색	5YR 3/6	chestnut brown
풀색	5GY 5/8	grass green	분홍	10RP 7/8	pink
초콜릿색	5YR 2/2	chocolate	쑥색	5GY 4/4	rtemisia(green)
로즈핑크	10RP 7/8	rose pink	계란색	7.5YR 8/4	eggshell
올리브그린	5GY 3/4	olive green	포도주색	10RP 2/8	wine
귤색	7.5YR 7/14	tangerine	연두색	7.5GY 7/10	yellow green
하양 (흰색)	N9.5	white	호박색(광물)	7.5YR 6/10	amber

잔디색	7.5GY 5/8	lawn(green)	흰눈색	N9.25	white snow
가죽색	7.5YR 6/6	buff	대나무색	7.5GY 4/6	bamboo
은회색	N8.5	silver grey	캐러멜색	7.5YR 5/8	caramel
멜론색	10GY 8/6	melon	시멘트색	N6	cement
커피색	7.5YR 3/4	coffe	초록	2.5G 4/10	green
회색	N5	gray	흑갈	7.5YR 2/2	blackish brown
에메랄드그린	5G 5/8	emerald green	목탄색	N2	charcoal grey
진주색	10YR 9/1	pearl	옥색	7.5G 8/6	jade
검정(검은색)	N0.5	black			

CHAPTER 02

Name of Hairstyle and Description

Afro : Big hair hairstyle, natural comb-out, featured heavily in African-American culture, popular through the late 1960s and 1970s in the United States of America.

Asymmetric cut : A haircut where the hair is longer on one side. In the 1980' s and 1990' s, asymmetric were a popular staple of Black hip hop fashion, among women and men.

Beehive : Backcombing or teasing with hairspray to style hair on top of the head so that the size and shape is suggestive of a beehive, hence the name.

Bettie Page : Bangs (or fringe) straight across the high forehead, or cut at a slight U-shape.

Big hair : Any hairstyle with large volume, though this is generally a description given to hair with a straight texture that is blown out or "teased" into a large size. The increased volume is often maintained with the use of hairspray or other styling products that offer hold.

Blow Out : Hair that is used with rich products and blown dry from the roots to the ends.

Blunt cut : A women' s any-length hairstyle with the bottom of the hair cut straight across leaving no texture.

Bob cut : A classic short hairstyle where it is cut above the shoulders in a blunt cut with typically no layers. This style is most common among women.

Bouffant : A style characterized by smooth hair that is heightened and given extra fullness over teasing in the fringe area.

Bowl cut : Named for the shape of the style as much as for a once common method of achieving it (i.e. using a bowl to style the cut by placing it on the head and trimming the exposed hair). Moe Howard from The Three Stooges has this hairstyle for his trademark and Henry V of England had a similar hairstyle. This hairstyle was popular in America (for straight hair textures) during the late 1980s and early 1990s.

Braid : A braid or braids, also known as a plait or plaits or a tress or tresses, is a type of hairstyle usually worn by women with long hair in which all or part of one' s hair is separated into strands, normally three, and then plaited or braided together, typically forming one braid hanging down at the back of the head or two braids hanging down on either side of the head. Braids can also be strengthened or lengthened with the addition of weaving hair. Braids may also serve as the base for sewing on additional weave hair on top of the braid for additional hair styling options. See also cornrows, queue, crown braid (below), Dutch braid and pigtail.

Brush cut : Alternative name for a longer butch cut, especially one that shows natural curl.

Bun : A women' s hairstyle where the hair is pulled into a knot at the back or top of the head. It can also consist of a bun at the side of the head or two buns on either side of the head. See also chignon.

Bunches : Another name for pigtails worn braided or unbraided.

Burr : Alternative name for a butch that is less than one quarter of an inch on top.

Business-man cut : Alternative name for a regular haircut.

Butch cut : A butch is a type of haircut in which the hair on the top of the head is cut short in every dimension. The top and the upper portion of the back and sides are cut the same length, which generally ranges between one quarter of an inch and three quarters of an inch, following the contour of the head. The hair below the upper portion of the sides and back of the head is tapered short or semi-short with a clipper, in the same manner as a crew cut.

Buzz cut : A buzz cut is any of a variety of short hairstyles usually designed with electric clippers. Buzz cut styles include the butch cut, crew cut, flattop and ivy league.

Caesar cut : The Caesar cut is a men' s hairstyle which is cut to a regular fade with the bangs or fringe left longer than the top length.

Chignon : A loan word from French for this particular style of bun which is pinned at the nape of the neck and requires tighter binding than ordinary buns.

Chonmage : A traditional Japanese haircut.

Comb over : Hair that is combed from one part of the head to another often to cover up a bald spot.

Conk : A hairstyle where Afro-textured hair is straightened. This was popular among African American men from the 1920s to 1960s.

Cornrows : A hairstyle originating from Subsaharan Africa, popularized by African Americans wherein the hair is braided into a series of French-braid-like locks that cling to the head and travel the neck.

Crew cut : A crew cut is a type of haircut in which the hair on the top of the head is cut relatively short, graduated in length from the longest hair that forms a short pomp (pompadour) at the front hairline to the shortest at the back of the crown. The hair on the sides and back of the head is usually tapered short, semi-short or medium.

Cropped hair : A short hairstyle in which the hair is cut very close to the head, with the hair left long enough to either touch or go past the fringe.

Crown braid : A crown braid or crown plait is a traditional Ukrainian hairstyle usually worn by women with long hair. It consists of a single braid wrapped around the head, and is the trademark hairstyle worn by Ukrainian politician Yulia Tymoshenko. It is also similar to some of the hairstyles worn by Frida Kahlo. See also braid or plait.

Croydon facelift : A tight ponytail worn at the top of the head, giving the effect of a facelift as it tugs the facial skin upwards. Often inappropriately referred to as a "top knot" (though this term, in its original context, actually refers to a chonmage).

Curtained hair : Curtained hair is the term given to the hairstyle featuring a long fringe divided in either a middle parting or a side parting. This hairstyle was popular on adolescents and men during the late 1980s and early 1990s.

Devilock : The sides and back are kept short while the front is grown long and combed forward. Well known in certain punk rock subcultures due to its use being commonly associated to Misfits bassist Jerry Only.

Dice Bob : A six-sided haircut, in that the hair is shaped with various flat planes.

Dido flip : A "short choppy shag", popularized by British pop Dido

Dreadlocks : The hair is 'dreadlocked' into individual sections using one of several methods, usually either backcombing, braiding, hand rolling, or locking in the roots. Dreadlocks can vary in size, tightness, and length. Dreadlocks are associated most closely with Rastafarians.

Duck's ass or D.A. : A hairstyle predominantly favoured by men; though some young female fans of Elvis Presley sported a similar look in his heyday. The style required that the hair be combed back around the sides of the head. The teeth edge of a comb was then used to define a central parting running from the crown to the nape at the back of the head, resembling, to many, the rear end of a duck. The hair on the top front of the head was usually that of a pompadour. The sides on back were styled to resemble the folded wings of the duck.

Eton crop : A very short, slicked-down blunt crop.

Emo hair : Hair is often feathered by that of a razor cut, known as a "modern mullett" whereas the top is textured and shorter than the back/length. popularized by "Emos".

Fade : Alternative name for a short taper.

Fallera hairdo : A Valencian fallera hairstyle is a traditional women' s hairstyle typical of Valencia and its Falles celebration, and is reminiscent of the "Princess Leia hairstyle" of Star Wars fame. It is an elaborate hairstyle consisting of two spiralled buns on either side of the head and two braids wound around the back of the head, with twisted strands of hair in different figures in the middle of the braids. This hairstyle is reminiscent of the one found on the ancient statue of the Lady of Elx.

Fauxhawk : Also known as the 'frohawk' , this hairstyle is an approximation of a mohawk made without shaving or 'buzzing' the hair on the sides of the head, thus allowing an imitation of the look of a true mohawk without having to commit to removing most of one' s hair. The name is a portmanteau of faux, the French word for false or fake, and 'mohawk' .

Feathered hair : Feathered hair was popular in the 1970s and the early 1980s with both men and women. The hair was grown long on both sides (normally covering the ears, although it could be down to the shoulders), left unlayered (although some men with curly hair did have it layered) with either a side or a centre parting.

Finger wave : Hair is set into curls using lotion and a comb.

Fishtail hair : Hair is worn in a braid resembling a fish' s tail.

Flattop : A type of crew cut where the hair on the top of the head is cut as a flat plane giving a levelled 'flat-topped' look.

Flipped hair : A women' s style that was popular in the 1960s. Characterized by upward curling ("flipped") ends. Lesley Gore wore the flip in 1963 and Cybill Shepherd wore it as a beauty queen in 1966.

Fontange : A hairdo popular in the second half of the 17th century.

French braid : A French braid is a braid that appears to be braided "into" the hair, often described as braided backwards strands, going over instead of under as in a Dutch braid.

French twist : A hairstyle wherein the hair is twisted behind the head into a sort of bun style.

Fringe (bangs) : Hair that is combed to the front of the head and cut at or above the eyebrows.

Frosted tips : A men' s hairstyle (though women have adopted as well) in which the hair is cut short and formed into short spikes with hair gel or hair spray. The hair is bleached such that the tips of each spike will be pale blond, usually in contrast to the wearer' s main hair color. Frosted tips were prominent throughout the late 1990s.

Full crown : Alternative and historic name for a short taper.

G.I. haircut : Alternative name for a crew cut.

Hair extensions : A hairstyle that is lengthened by additions of human or synthetic hair.

Half crown : Alternative and historic name for a semi-short taper.

Half updo or Half Up, Half Down : Popularised in the 1960s by sex icons like Brigitte Bardot, this women' s hairstyle requires medium-length or longer hair. The hair is divided from the temples back and then fixed into bun, chignon or ponytail to secure into place (therefore, half the hair is in an 'updo'). If a woman has a fringe, that area is usually left free and not pulled into the 'updo' . Frequently, the hair at the crown of the head is "teased" or backcombed first to create volume.

Harvard clip : Alternative name for an Ivy League.

High and tight : A military variant of the crew cut.

Highlights : The style involves highlighting or dying various thin sections of hair through hair coloring.

Hime cut : A hairstyle that consists of long straight hair that reaches at least below the shoulder blades with part of it cut to about shoulder-length and a fringe (bangs) that reaches the eyebrows.

Historical Christian hairstyles : Hairstyles worn by Christians in times past.

Hi-top fade : The hair is cut short on the sides and is grown long on the top. This style was popular among African-American youth and men in the late 1980s and early 1990s.

Induction cut : A haircut given to recruits being inducted into military service.

Ivy League : An Ivy League, also known as a Harvard Clip or Princeton, is a type of crew cut in which the hair on the top of the head is long enough to style with a side part.

Jewfro : A Jewish variant of the 'Afro' hairdo.

Jheri curl : Hair that is curly and kept moist (or at least a wet-look maintained) by a Jheri curl activator. This style was popular in the 1980s especially with African-Americans, including Michael Jackson and Rick James.

Layered hair : A women' s hairstyle where different sections of the front hair are cut at different lengths to give the impression of layers.

Liberty spikes : Hair that is grown out long and spiked up usually with a gel.

Line up : Hair that has an even line across the forehead and then turns sharply at a 90 degree angle and blends with the sideburns. It can be worn with almost any other hairstyle. Primarily worn by Hispanics and some African Americans. Famous examples include hip hop artist Drake (entertainer) and Vinny Guadagnino from Jersey Shore.

Long hair : Hair that is grown long and flowing freely.

Marcelling : Deep waves made in hair by a heated curling iron.

Mod cut : A men' s haircut where the hair flips outward, like wings.

Mohawk : Hair that is shaved or buzzed on the sides leaving a strip of hair in the middle. It is often spiked up.

Mop-Top : A mid-length haircut that has a fringe (bangs) that brushes over the forehead, collar length at back, with the ears partly covered by the hair, dependent on style. Was made popular by The Beatles (John Lennon, Paul McCartney, George Harrison, and Ringo Starr)

Mullet : Hair that is short in front and long in the back. Often described as "Business in the front, party in the back".

Odango : A hairstyle where two buns are worn on either side of the head, with the rest of the hair worn as pigtails. Most well-known example of this style is worn by the title character of the Japanese anime Sailor Moon. In African American culture, commonly referred to as "Afro Puffs."

Oseledets : A haircut consisting of a long tuft or lock of hair left on top or on the back of the shaven head of a male Orthodox Hindu. The style is also associated with Cossacks.

Pageboy : A longer version of a bob, typically worn with a fringe (bangs) and reaching shoulder-length or a little longer.

Part : Hair that is parted.

Payot : Sidelocks or earlocks, worn by some males in the Orthodox Jewish community.

Perm : Hair that is chemically treated to retain curl (curly perm) or lie straight (straight perm). New hair is not affected by the chemical treatment and will grow out naturally.

Pigtails : Hair that is parted into sections or into just one section; that is a smaller or shorter version of a ponytail. Pigtails are short and end with a curl similar to a pigs' tails.

Pixie cut : A short wispy women' s hairstyle with or without a shaggy fringe (bangs).

Pompadour : The hair is swept upwards from the face and worn high over the forehead, and sometimes upswept around the sides and back as well. The style, named for Madame de Pompadour (1721 1764, mistress of King Louis XV), is for both women and men.

Ponyhawk : 1. A combination of a mohawk with a ponytail in the back. 2. Longer hair worn in several ponytails running from front of the head to the back of the head resembling a mohawk. Unlike normal mohawk, hair is not cut from the sides. The word is a portmanteau of 'ponytail' and 'mohawk'.

Ponytail : hair that is pulled into one or more sections held in a rubber-band(s) or by ribbon(s) anywhere on a head that is similar to a horse's or pony's tail. Ponytails are longer versions of pigtails. The length of a ponytail may be an entire curl, may end with a curl, or may be entirely straight without a single curl, whereas pigtails end with curls similar to a pig's tail.

Princeton : Alternative name for an "Ivy League" haircut.

Professional cut : Alternative name for a regular haircut.

Psychobilly Wedge : The Psychobilly Wedge is a sort of mix between a mohawk hairstyle and the pompadour. The hair along the side of the head is shaved and the middle is not spiked but slicked back and fashioned in a pompadour.

Queue : Hair is worn long and gathered up into a ponytail, often braided. It was worn traditionally by the Manchu people of Manchuria and certain Native American groups.

Quiff : The quiff combines the 1950s pompadour hairstyle, the 1950s flat-top, and, sometimes a mohawk. The hairstyle was a staple in the British 'Teddy Boy' movement, and became popular again in Europe in the early 1980s and

2010s. Recently examples of people sporting quiffs are Alex Turner and Matt Helders of Arctic Monkeys, Tom Meighan of Kasabian, Eugene McGuinness, Bruno Mars, Nick Grimshaw, and One Direction.

The Rachel : A square-layered haircut popularized by Jennifer Aniston as Rachel Green on the Friends sitcom.

Rattail : Hair that has been grown out long and shorn, or 'buzzed' except for a long, thin braid at the very top of the neck.

Razor cut : Includes choppy short layers, thinned at the bottom. Not dissimilar from "emo" hair. Usually has a side-sweep fringe.

Recon : A radical version of the high and tight with the sides and back cleanly shorn very high up the head, intentionally leaving a very extreme contrast between the longer top hair and the shorn sides.

Regular haircut : A regular haircut is a men's and boy's hairstyle that has combable length on top, a defined or deconstructed side part, and a short, semi-short, medium, long, or extra long back and sides. The style is also known by other names including taper cut, regular taper cut, side-part and standard haircut; as well as short back and sides, business-man cut and professional cut, subject to varying national, regional, and local interpretations of the specific taper for the back and sides. Important to note that there is an underlying assumption of "regular" as being male, and most often "straight texture."

Regular taper cut : Alternative name for a regular haircut.

Ringlets : A tightly curled hairstyle.

Shag : A choppy layered hairstyle, characterized by layers to create fullness in the crown and fringes around the edges. There are many versions including the frat shag and boy' s shag.

Shingle bob : A dome-shaped short haircut with the bottom razored into a V shape.

Short back and sides : Alternative name for a regular haircut.

Short brush cut : Alternative name for a butch cut, especially one that shows natural curl.

Short hair : Hair that is cut short. "Short" is subjective given personal, cultural, historical, and generational perspectives. It may indicate hair shorter than chin length.

Slicked-back : Hair that is combed back, away from the forehead, then held in place with a hairstyling product.

Spiky hair : All hair sticking up, like spikes sticking out from the head.

Standard haircut : Alternative name for a regular haircut.

Surfer hair : A tousled hairstyle.

Taper cut : Alternative name for a regular haircut.

Tail on back : Boys' hair style by growing the hair in back like a small tail. This is widely used all over India.

Tonsure : Traditionally worn by monks in the Middle Ages, still worn by some traditional monks today.

Undercut : The undercut is a unisex haircut whereby the top section of hair is held in place whilst the side and back sections are cut, thus making the top longer and the back and sides undercutting. See also bowl cut.

Updo : An updo is the hairstyle in which the hair is twisted or pulled up.

Waves : Short Hair Waves, shortened to just waves, is a very common and sought after hairstyle for African American men that create the appearance of water like waves of the hair.

Wings : Typically long, the style can range from long and drooping below the eyes, to a shorter length. They are wavy and if straight, the length comes to halfway down the ears. Instead of lying on the wearer' s ears, the hair flips up and comes straight out like an airplane wing, hence the name. Popular amongst the skateboarding, surfer and preppy communities

CHAPTER 03

미용관련 용어

A

abdomen : 복부

absorb : 흡수하다

accent : 강조

according to : ..에 따르면, ..에 의하면

accutane : 여드름 약의 이름

achromatic color : 무채색

acne : 여드름 (pimple, comedo)

acne spot : 여드름 자국

acrylic bowl : 아크릴 보울

acrylic brush : 아크릴 브러쉬

acrylic liquid : 아크릴 액

acrylic powder : 아크릴 가루

actually : 실제로, 정말로

AD : anno Domini, 서기

adapt : 맞추다. 적응하다.

additional : 추가적인

adhesive : 접착제

adjust : 조정하다, 적응하다.

adsorption : 흡착

advance : 진전, 발전

advanced : 고급의

advice : 충고, 조언

after shampoo : 에프터 샴푸

aftershave lotion : 에프터쉐이브 로션

age spot : 검버섯

aging skin : 노화피부

agitated : 뒤흔들다

agree with : ...에 동의하다

AIDs : 후천성 면역결핍증

air brush : 에어 브러쉬

allergic reaction : 알레르기 반응

allow : 허락하다, 용납하다

almond shape : 아몬드형

along : ...을 따라서

am : 오전, ante meridiem의 약자

amazing : 놀라운

ambergris : 향

ancient : 고대의

anniversary : 기념일

antiseptic spray : 소독제 분무기

apologize : 사과하다.

apply .A.. to B : A를 B에 바르다

appointment : 임명, 약속.

apron : 에이프런

area : 지역, 부분

aroma : 향 또는 향기

aroma oil : 아로마 오일

aromatherapy : 아로마테라피

assist : 돕다

assistant : 보조

assume : (책임) 맡다, ... responsibility for ...에 대해 책임을 지다

asymmetry : 비대칭
at home= comfortable : 편안하게
athlete' s foot : 무좀
attend : 참여하다
attention : 주의
attire : 의복, 복장
attractive : 매력적인
attractiveness : 매력, 애교
autograph : (연예인) 사인
availability : 이용도, 이용가능성
available : 이용할 수 있는, 쓸모 있는, 유효한
avoid : 방지하다, 막다. 피하다.
avoiding : 회피

B

base : 베이스
base coat : 베이스 코트
bath products : 목욕용 제품
BC : 기원 전, before Christ의 약자
beard : 수염
beating : 가볍게 주먹을 쥐어 두드리기
begin : 시작하다
benefit : 혜택, 이득
benefit : 혜택, 이득
bet : 내기, 생각
bill : 계산서, 청구서
black head : 흑면포
bleach : 표백, 탈색
bleach hair : 탈색 모발
blemish : 잡티
blend : 섞다. 혼합하다.
blocking : 브로킹
blood : 혈액
blood vessels : 혈관
blow dry : 말려서 드라이하다.
blunt cut : 블런트 커트
blushing : 얼굴을 붉힘, 얼굴이 빨개짐
body cream : 바디크림
body lotion : 바디로션
body of nail plate : 조체
body product : 바디제품
body skin : 전신피부
book: [명] 책, [동] 예약하다
bowl : [명] 보울
bracelet : [명] 팔찌
bride : [명] 신부
bridegroom : [명] 신랑
broken: [형] 깨진, 부러진
brown hue: 갈색조
brownish brunette: 갈색 빛나는 검은색
brunette: 검은색
brush cleaner : 브러쉬 크리너
bush: 관목, 덤불

C

camel brown : 담황갈색
cancel : 취소하다.
capillary : 모세관
capillary vessels : 모세혈관
carefully : 조심스럽게, 주의해서
carrier oil : 캐리어 오일
celebrate : 축하, 경축하다.
cellulite : 셀룰라이트, 귤껍질모양의 피부
ceremony : 의식 식 격식
check : 살피다.
cheerful : 발랄한, 쾌적한
chemical : 화학적인
chemical service : 화학적인 시술(퍼머, 염색)
chestnut brown : 밤색
chin : 턱
chip : 조작, 부스러기
choice : 선택, 선택권
chromatic color : 유채색
chrome : 채도
chucking : 가볍게 치듯이 문지르기
circulation : 순환, 유통
circulatory system : 순환계
circumstance : 상황, 환경
clarifying : 명확하게 하다. 투명하게 만들다.
classic : 고전적인, 고상한
claustrophobia : 폐쇄공포증
clean up : 깨끗이 하다
cleansing : 세정
cleansing gel : 크린싱 젤
cleansing oil : 크린싱 오일
cleansing water : 크린싱 워터
clear or lucid layer : 투명층
clipper : 클리퍼, 바리캉(barican)
clipping : 클리핑
close comedo : 백면포, = white comedo
coloring : 염색하다.
coloring bowl : 컬리링 볼
coloring hair : 염색 모발
coloring pigments : 착색안료
coloring wagon : 이동식 염색정리대
comb : 빗
comb out : 콤 아웃
combination : 잘 어울리는 배합
combination skin : 혼합피부
come off : 떼어 낼 수 있다, 제거 할 수 있다
comedo(nes) : 여드름
comedo : 면포 단계의 여드름
comfortable : 편안한
complement : 보충하다.
complexion : 안색, 양상
comprehensive : 포괄적인
compress : 습포법
compression : 압박법(누르기)
concern : (동) 영향을 미치다
congratulations : 축하
consider : 여기다, 고려하다
consist of : ...로 구성되다
consultation : 상담
consulting : 상담
contrast : 대조
convenient : 편리한
corneous tissue : 각질조직

correct : 맞는, 정확한, 옳은
corrugation : 주름 잡힌 손톱
cortex : 모피질
cosmetic surgery : 성형외과
counseling : 상담
cowlick : 뻣뻣이 들고 일어나는 머리칼
cracked : 금이 간, 갈라진
creating : 창조하다. 자아내다.
credit : 신용의
croquignole winding : 크로키놀식 와인딩, 크로키놀식 감기
cupping : 손을 진공상태(컵 모양)로 하여 두드리기
cure : 관리
curvature : 만곡
curve : 커브
customer card : 고객카드 작성
customer : (가게의) 손님, 고객
cut off : 잘라내다.
cuticle : 모표피, 거스러미 (손톱뿌리를 덮은), 얇은 피부
cuticle oil : 큐티클 오일
cuticle remover : 큐티클 리무버
cuticle softener : 큐티클 소프트너
cutting scissors : 커트 가위

D

daily : 매일, 나날의
damage : 손상
damaged hair : 손상된 머리
dandruff : 비듬
dangle : 매달리다. 달랑거리다
dark chocolate : 진한 초콜릿 색
day cream : 데이 크림
dead skin cell : 각질세포 (horny substance, keratin)
decide: 결정하다
deep cleansing : 딥 크린싱
deep stroking : 심부 무찰법
deluxe : 호화스러운, 사치스러운, 고급의
departure : 출발
depending on : ...에 따라
depilation : 제모
deposit : 예약금
deposit : 착수금, 보증금
depressor anguli oris : 구각하체근
depressor libii inferioris : 하순하체근
dermal papilla : 진피유두
dermal root sheath : 진피근초
dermatitis : 피부염
dermatologic immunology : 피부 면역학
dermis : 진피
desincrustation : 각질제거
despite : ...에도 불구하고
detail : 상세, 세부
diabetes : 당뇨병
diagnosis : 진단
decide : 결정하다.
dietary manipulation : 식이 조작
difficulty : 어려움, 곤경, 장애
dig : 파다
dilatation : 확장
dilate : 확장시키다,

direction : 방향
discharge : 빼내다, 흘러 나오게하다(고름)
discomfort : 불편, 불편하게 하는 것
discord : 비조화
distend : 팽창하다
divide : 나누다, 분열하다
dominance : 강조
double wrap : 이중 랩
draw : 그리다, 끌어 당기다
drip on : ....에 떨어지다
dry cut : 드라이 커트
dry skin : 건성피부
drying : 말리기, 건조
dull : 둔탁한, 칙칙한
duration : 시간
dye : 염료, 염색하다.
dye cape : 염색보

E

earring : 귀걸이
ease : 편안함, 안락함
easy to manage : 관리하기 쉬운.
edema : 부종
effleurage : 경찰법(쓰다듬기, stroking)
elasticity : 신축성, 탄력성
electrical : 전기의
elegant : 우아한
emery board : 에머리 보드
emollient & moisturizing function : 유연 및 보습 작용
enamel dryer : 에나멜 건조기
enamel marbling : 에나멜 마블링
enamel remover : 에나멜 리무버
end taper : 엔드 페이퍼
enough : 충분히, 충분한
ensure: 반드시 …하게 하다, 보장하다
enzyme : 효소
eponychium : 에포니키움
essential oil : 에센셜 오일
establishment : 영업장소, 사업장
even : (형) 평평한, 고른, (명) 걱정, 근심
exclusively : 독점적으로, 오로지
excoriation : 찰창
exfoliate : 벗겨내다. 박리시키다.
exfoliation : 박리, 벗겨짐
expression : 압착, 표현
extension : 확장
extract: 짜내다
extraction: 추출, 적출
eyebrow: 눈썹
eyelash : 속눈썹

F

fabric wrap : 훼브릭 랩
fabulous : 훌륭한, 멋진
face powder : 분
facial : 얼굴의
facilities : 시설
fascinating : 대단히 흥미로운, 매력적인

fatigue: 피로
favorite: 마음에 드는, 매우 좋아하는
feathering : 페더링
feel: 느끼다
fiery: 불같은
figure: 모양, 체격, 형태
file : 파일
filler : 필러
filler powder : 필러 파우더
film agent : 막형성제
finger bowl : 핑거 볼
finger grip : 약지환
finish up : 끝내다
first rinsing : 1차 헹구기
fix : 고치다
flattered : 추켜세워진, 칭찬의 말 때문에 부끄러운
flexibility : 신축성 있는, 잘 구부러지는, 유연성
flip out : ...이 밖으로 뻗치다
flow : 흐르다
focus : 집중하다. 맞추다
foot powder : 발 전용 파우더
formal : 격식을 차린, 정중한
foundation : 파운데이션
free edge : 자유연(손톱 끝)
french manicure : 후렌치 매니큐어
frequency : 빈도, 빈발, 잦음
friction : 강찰법, 마찰법(문지르기)
front : 앞
front desk : 프런트, 접수계
fruits : 과일
full coat : 가득 칠하다
furrow : 고랑 파진 손톱

G

gel : 젤
gel system : 젤 시스템
general shampoo : 일반 샴푸
gentle : 온화한, 조심스러운
get off : 내리다
glam/glamour/glamorous : 글래머
gloss/glossy : 윤이 나는
glossy : 반짝거리는, (머릿결이) 매끄러운
glue : 글루, 풀
go ahead : 계속하다
go down : 가라앉다
go well with : ...와 잘 어울리다.
golden brown : 금갈색
gorgeous : 아주 멋진, 눈부신, 훌륭한
gradation : 그라데이션
guaranteed : 확보된, 보증된
guide book : 안내 책자

H

hacking : 손의 측면 또는 손등으로 두드리기
hair : 모발
hair band : 헤어밴드
hair cream : 헤어 크림
hair diagnosis : 모발 진단
hair follicle : 모낭
hair piece: 부분 가발
hair root : 모근부
hair shaft : 모간부

hair shaping : 헤어 셰이핑
hair treatment : 헤어 트리트먼트
half moon : 반월(루눌라)
half off base : 하프 오프 베이스
hand cream : 핸드크림
hand lotion : 핸드로션
hand paint : 핸드 페인트
hang nail : 행 네일(거스러미 손톱)
hang on : 잠시 기다리다 (전화를 끊지 않고 두다)
hard wax : 하드왁스
healing : 치유, 치료
health challenge: 건강 위해요소
healthier : 건강한
heart problems : 심장병
heat : 열, 열기
heat wave : 열 펴머, 히트 웨이브
heaviness : 무거움, 중압감
help out : 거들다. 도와주다.
herpes simplex : 단순 포진
hesitate : 망설이다, 꺼리끼다
high blood pressure : 고혈압
home hair care advice : 홈케어, 조언
hope : 희망하다. 바라다
hormone imbalance : 호르몬 이상
horney layer : 각질층
horse' tail : 호스 테일, 말꼬리
hot compress : 온습포
hot oil manicure : 핫 오일 매니큐어
hue : 색상
hurt : 다치게 하다, 아프게 하다
hygiene : 위생
hyponychium : 하이포니키움, 하조피

I

important : 중요한
improve : 개선하다
include : 포함하다
index : 색인
infection : 감염
ingredient : 성분, 원료, 재료
ingrown nail : 인그로운 네일(속으로 자라는 손톱)
inhalation : 흡입법
introduce : 소개하다
irritate : 자극하다
isadora : 이사도라
itchy : 가려운

J

jewelry : 보석류
jojoba oil : 호호바 오일

K

kit : 장비 일습
kneading : 유날법(반죽하기)

L

last : 마지막
layer : 층
lean on : ...에 기대다.
leather : 가죽

length : 길이
leukonychia : 루코니키아(백색 조갑)
lie down : 눕다
lie on : ...에 눕다.
lifting : 리프팅
ligament : 인대
light : 가벼운, 옅은
light purple: 옅은 자주색
light tan: 옅은 황토색
lip stick : 립스틱
liquid wrap : 리퀴드 랩
long stroke : 롱 스트록
look good on : 잘 어울리다. .
loose powder : 가루분
loose : 헐거운, 느슨한, 풀린
loosen : 느슨하게 하다. 풀다
low class : 하류층
lrregular bone : 불규칙골
lukewarm : 미지근한
lunula : 루눌라, 반월
lymph : 림프
lymph node : 림프절
lymphatic vessel : 림프관

M

maintain : 관리하다, 유지하다
major : 대부분의, 주요한
makeup remover : 화장 지우기 화장품
manager : 지배인, 관리자
manicure glue : 매니큐어 글루
manicurist : 매니큐어리스트
marble : 대리석
marbling : 마블링
mark : 점 (beauty mark, birth mark)
massage cream : 마사지 크림
match : 성냥, 경기, ...에 맞추다.
meal : 식사
media : 매개체
medieval : 중세의
medium length : 중간 길이
medium stroke : 미디움 스트록
messy : 지저분한, 엉망인
metal pusher : 메탈 푸셔
micro file : 마이크로 파일
mini scissors : 미니 가위
miniature : 축소모형
minimize : 최소화하다, 축소하다
misplace : ...을 잘못 두다
miss : 빠뜨리다
moisture : 습기, 수분
moisturizing cream : 보습크림
moisturizing essence : 보습 엣센스
motion : 동작
movement : 운동
mud : 머드
muscle: 근육

N

nail : 발톱
nail artist : 네일 아티스트
nail bar : 네일 바
nail bed : 네일 베드(조상)

nail bleach : 네일 블리치
nail body : 네일 바디
nail brush : 네일 브러쉬
nail clipper : 네일 클리퍼
nail enamel : 네일 에나멜
nail enamel thinner : 네일 에나멜 희석제
nail fold : 네일 폴드(조주름)
nail form : 네일 폼
nail groove : 네일구르브
nail lacquer : 네일 락카
nail matrix : 매트릭스
nail polish : 네일 팔리쉬
nail salon : 네일 살롱
nail strengthener : 네일 보강제
nail tip : 네일 팁
nail wall : 네일 월(조벽)
nailist : 네일리스트
nape : 목뒤, 네이프
natural color : 자연 색
necklace: 목걸이
nervous: 불안 해 하는, 신경이 과민한
neutralizer: 중화제
night cream : 나이트 크림
nipper : 니퍼
nobility: 고귀함, 고결함
non acetone : 비아세톤
normal skin : 정상피부
normal taper : 노말 테이퍼
notching cut : 나칭 커트
nourish : 자양분을 주다
nourishing cream : 영양 크림
nourishment: 영양분, 자양분
now a day : 요즘의, 오늘날의
nutrition : 영양
nutrition supply : 영양 공급

O

oblique : 완곡한, 사선의
oblong : 장방형의
occupation : 직업, 업무, 심심풀이
oily : 유성의
oily skin : 지성피부
omit : 빠뜨리다. 생략하다
ompare to : …와 비교하다
one length : 원랭스
onychatrophia : 조갑 위축
onychia : 오니키아
open comedo : 흑면포, = black comedo
opening time : 영업 시작 시간
opinion : 의견, 견해
ordinary razor : 오디너리 레져(일상용 레져)
origin : 기시
originate : 유래하다
other than : … 말고, …과는 달리
oval shape : 타원형, 계란형
over processing : 오버 프로세싱
oxidation : 산화작용

P

packs & masks : 팩 & 마스크류
pain : 통증
paper towel : 페이퍼 타올

paper wrap : 페이퍼 랩
papule : 구진
paraffin manicure : 파라핀 매니큐어
paraffin wax : 파라핀 왁스
parallel bob : 평행보브
parking lot : (미) 주차장
particular : 특별한, 특정의
past : ...를 지나서, 지난
path : 통로, 길
patience: 인내, 참을성
patting : 고타법
payment : 지불, 계산
peach color : 복숭아 색
pedi file : 페디 파일
pedi stone : 페디 스톤
penetrate : 뚫고 들어가다, 관통하다
perfect choice : 탁월한 선택
perfume : 향수
period : 시대, 시기
perm rod : (퍼머용) 작은 막대기
perm solution : 퍼머 액
permanent coloring : 영구적 염색
permanent wave : 퍼머넌트 웨이브
personal belongings : 개인 소지품
petrissage : 유연법(집어주기)
petrissage : 집어주기
pic : 사진, picture의 약자
pigmentation : 착색, 염색
pimple : 여드름
pinching : 압박법
pink color : 핑크색
place : 장소, 곳
plain shampoo : 플레인 샴푸
plastic surgery : 성형외과
pleasure : 기쁨, 만족
plump : 통통하게 살이 찌다
pm : 오후, post meridiem의 약자
point shape : 포인트 쉐입
pony tail : 포니 테일
poodle cut : 푸들 컷
popular : 유행하는
pore : 구멍
porous : 부서지기 쉬운, 다공성의, brittle, fragile
portrait : 인물사진, 흉상
position : 자세
prefer : ....을 더 좋아하다, 선호하다
prepare : 준비하다.
preservative : 방부제
press : 누르다
pressure : 압력
pressure shaking : 압박 진동법
pretty : 아주, 매우, 매력적인, 예쁜
previous : 이전의, 먼저
primer : 프라이머
privilege: 특전, 특권
privilege: 특전, 특권
procedure : 순서, 진행
proceed : , 계속해 나가다
product : 물품
progression : 진행
promote : 촉진시키다
proper : 올바른, 적절한
prospective : 예상된, 장차의
protect : 보호하다.

protection : 보호
provide : 제공하다, 규정하다
psoriasis : 마른 버짐, 건선
pulling : 당기면서 주무르기
pulling shaking : 견인 진동법(끌어당기면서 흔들기)
purchase : 구입, 구매
pus : 고름, 농
pusher : 푸셔
pustule : 농포
put down : 적어 놓다

Q

quality : 품질, 양질
quare : 네모 각진
quasi permanent coloring : 반영구적 염색
quats : 화합물
quick dry : 빠른 건조
quickly: 빨리

R

raking : 긁어주기
rash : 발진
rate & rhythm : 속도와 리듬
razor : 레자
really : 정말로, 진심으로
recently : 최근에
recommend : 추천하다
rectangular (square) shape : 직사각형,네모형
refer to : ..와 관련이 있다
reflexology : 반사학
regular manicure : 레귤러 매니큐어
rejuvenate : 다시 젊어 보이게 하다, 활기를 되찾게 하다
relax : 이완, 안심
relieve : 완화시키다. 안도하게 하다
remainder : 나머지
remember : 기억하다.
reminding : 상기시키다
removal : 제거
repetition : 반복
request : 요청, 요구
require : 필요하다, 요구하다
retain : 보유하다
rewarding : 가치 있는
rhine stone : 라인 스톤
rid : 제거하다. 자유롭게 하다
rinse : 린스
rip : 찢다. 떼어내다.
riverse scissors : 리버스 가위
rod : 롯드
rod off : 롯드 박리
roll : 감다
rolling : 굴리기
root of nail : 네일 루트(조근)
rough : 거친
round: 원형의, 둥근
round shape : 둥근형
round square : 둥근 사각형
royalty : 귀족의
rub : 문지르다. 맞비비다
ruin : 몰락, 폐허, 망치다

S

safety : 안전, 안전성
safety razor : 세이프티 레이저
sanding block : 샌딩 블록
satisfied : 만족하는, 만족스러워 하는
satisfy : 만족시키다. 충족시키다
saturation : 포화상태
scaling : 세정
scalp : 두피
scar : 반흔
scent : 향기 향내
scissors : 가위
scrub : 문지르다
sculpture nail : 스켑춰 네일
seal : 봉하다, 밀봉하다
sebum : 피지
second rinsing : 2차 헹구기
secretion : 분비, 분비물
secure : 안전한, 확실한
seldom : 좀처럼, 드물게
selection : 선택, 선발, 선정
self-esteem : 자부심, 자존감
self-satisfaction : 자아만족감, 자기만족감
semi permanent coloring : 반영구 염색
senior citizen : 노인
sensitive : 민감한, 예민한
sensitive skin : 민감성 피부
serious : 심각한, 진지한
set up : 배정하다. 설정하다.
shade : 그늘, 빛 가리개
shampoo : 샴푸
shampoo solution : 샴푸액
shank : 다리
shape : 모양
shaping razor : 세이핑 레이저
shaving : 면도
shaving cream : 면도크림
sheen : 윤기, 광택
shiny : 빛나는, 반짝거리는
short stroke : 쇼트 스트록
shoulder : 어깨
shower gel : 샤워 젤
side : 사이드
sign : 서명, 간판, 표시, 신호
signature : 편지 말미 자필 서명
silk scissors : 실크 가위
skinny : 마른 여윈
skip : 건너 뛰다.
sleek : 매끄러운
slender : 날씬한, 호리호리한
slight : 약간의, 경미한
slim : 날씬한, 호리호리한
slim line: 가느다란
slipper : 슬리퍼
slithering : 슬리더닝
slough off : 벗겨지다.
smear : 마구 바르다
smell : 냄새가 나다
smooth : 매끄러운
smoothly : 부드럽게, 순조롭게
soak : 담그다, 흠뻑 적시다
soap : 비누
social status: 사회적 지위, 신분

sole : 발바닥, 단하나의, 유일한
sophisticated : 교양있는, 수준있는, 지적인
spaniel : 스파니엘
spatula : 주걱, 스파출라
special care : 특수 관리
special shampoo : 특수 샴푸
special treatment product: 특별 관리 제품
specialist : 전문가
spiral : 나선형의
spiral winding : 스파이럴 와인딩
splash: 떨어지다.
split ends : 갈라진 끝
spray gun : 스프레이 건
spread : 퍼지다, 펼치다
square: 네모난, 사각형
square shape : 스퀘어, 사각형
stain : 염색
stand ou t: 돋보이다, 특출나다.
start : 시작하다, 출발하다
status : 신분
stay in : 유지하다, 남아있다.
steam : 스팀, 수증기
steam distillation : 수증기 증류법
steamer: 스티머
stick : 봉, 막대기, 지팡이
stiffness: 뻣뻣한, 빽빽한, 근육이 결리는
stimulate : 자극하다.
stimulation : 촉진, 자극, 흥분
stone pusher : 스톤 푸셔
store : 보관하다. 저장하다.
straw blonde : 밝은 담황색
strengthen : 힘, 내구성, 견고력
strengthening : 보강
stripe : 줄무늬
stroke cut : 스트록 커트
stroke down: 쓸어내리다.
structure : 구조
styling : 스타일링, 스타일 만들기
subdivide : 다시 나누다, 세분하다
subtle : 미묘한, 절묘한
suffering : 고통, 괴로움
suitable : ...에 적합한
sun blocker : 햇빛 차단제
superfical stroking : 표면 무찰법(표면 쓰다듬기)
suppleness : 유연, 나긋나긋함
support : 지지
supporting : 뒷받침하는, 지탱하는
surfactant : 계면활성제
sustained pressure : 지속 압박법
suture : 봉합
swab brush : 컬러링 브러시 이름
swatch : 견본
sweet almond : 스위트아몬드
swelling : 부종
swimsuit : 수영복
swollen : 부은 상태의
symbol : 상징, 표시
symmetry : 대칭

T

T. P. O.: time, place, occasion
take off : 벗다

tapotement or percussion : 고타법(두드리기)
tapping : 손가락으로 피아노 치듯이 두드리기
technical : 기술적, 전문적인
temporary hair coloring : 일시적 염색
tension : 긴장, 긴장상태
terminal hair : 경모
test curl : 테스트 컬
texturizing scissors : 텍스쳐라이징 가위
therapy : 치료
thin : 얇은
thinning scissors : 틴닝 가위
thumb grip : 엄지환
tip : 붓끝
tip cutter : 팁 커터
tip overlay : 네일 팁을 이용한 아크릴릭 네일
toe separator : 토우 세퍼레이터 (발가락 띄우개)
toenails: 발톱
top coat : 탑 코트
towel dry: 수건으로 머리를 닦다. 말리다.
toxicity : 독성
transparent: 투명한, 명료한
treatment : 시술, 처치, 트리트먼트
treatment shampoo : 트리트먼트 샴푸
trim: 손질, 정돈.
troubled nails: 문제성 네일
tubercle : 결절
tuxedo: 턱시도
tweezer : 족집게

U

underneath: 아래의, 아래에
unfortunately: 불행하게도, 유감스럽게도
up style : 업 스타일
upper class: 상류층
upward: 위쪽을 향한

V

valuables : 귀중품
value : 가치
vanishing : 배니싱
vaporization : 확산법
varicose veins : 정맥류
various : 다양한, 여러 가지의
vibration : 진동법(떨어주기)
view : 견해 관점
vinyl cap : 비닐캡
virgin hair : 자연 모발
volume : 볼륨
volume : 부피, 분량
voluminous : 방대한, 넉넉한, 부피가 큰, 볼륨감 있는

W

warmer : 워머
worry : 걱정하다 걱정하게 만들다
washing agent : 세정제
wastes : 폐수, 불순물, 폐기물
water decals : 워터 디칼스

water marbling : 워터 마블링
water spray : 분무기
wavy : 곱슬거리는, 컬이 있는
wax : 왁스
waxing : 왁싱, 제모
wear a hat : 모자를 쓰다.
weaving comb : 위빙콤
wedding ceremony : 결혼식
wet cut : 습식 커트
white head : 백면포
whitening cosmetics : 미백 화장품
whole : 전체의, 모든, 온전한
wind : 말다, 감다.
winding : 와인딩, 감기, 말기
wipe : 닦다
wiry : 강단 있는, 뻣뻣한
worry: 걱정하다, 걱정하게 만들다
wrap : 랩
wrap plus : 랩 플러스
wringing : 짜듯이 비틀기
wrinkle : 주름

Y

year-end party : 망년회

[개정판]
미용인을 위한 실무영어

2011년 3월 15일 초판 발행
2016년 9월 1일 개정판 발행

저 자 김기영/신규옥/안경민
김수미/송다해/최은미

발행인 주 병 오

발행처 메디시인
경기도 파주시 회동길 209
파주출판문화정보산업단지
영업부 (031) 955-7566 · 7577
편집부 (031) 955-7731
FAX (031) 955-7730

ISBN 979-11-86739-25-9 정가 20,000 원